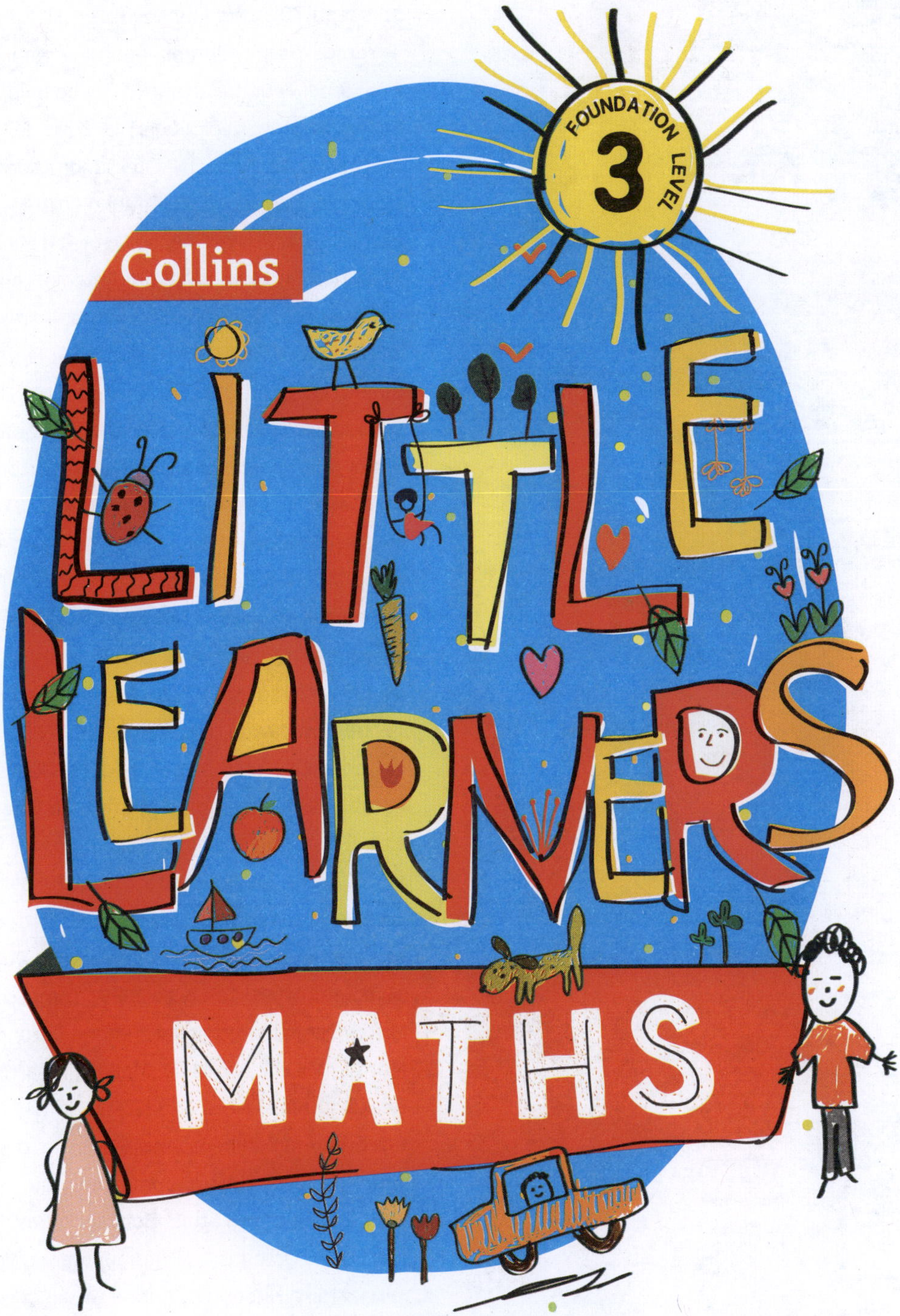

SERIES EDITOR: RANI T.

Published by Collins Learning
A division of HarperCollins*Publishers*
India Private Limited

HarperCollins Publishers
Building No. 10, Tower A, 4th Floor, DLF Cyber City, Gurugram, Haryana 122002, India

1 London Bridge Street, London, SE1 9GF, United Kingdom

2 Bloor Street East, Toronto, Ontario M4W 1A8, Canada

Lvl 13, 201 Elizabeth Street (PO Box A565, NSW, 1235), Sydney NSW 2000, Australia

195 Broadway, New York, NY 10007, USA

Browse the complete Collins catalogue at www.collins.in
For any queries/feedback, please write to us at **SCHOOLSUPPORT@harpercollins.co.in** or call us on **0120-4044800.**

First published in 2020

Reprint edition: 2025

ISBN: 978-935357-035-4

Artwork and layout by
Qualibre Creatives Pvt. Ltd.

Printed and bound by
MicroPrints India, New Delhi

Acknowledgements

Cover: HarperCollins*Publishers*India

Title Page: HarperCollins*Publishers*India

Prelim Pages: radenmas, denkcreative, nonohana, graphic-bee.

Inside: Pages 4: denkcreative; **5:** graphic-bee; **8:** MicrovOne; **11:** saenal78, Nina_Piankova; **13:** ONYXprj, Kittisak_ Taramas, Sergey Shambulin; pg **14:** KanKhem, Shendart, S-S-S; **15:** Tetiana Lazunova; **16:** bennyb, Denis08131; **17:** Ilyabolotov, Neokryuger; **18:** Zoya_Miller, Nadzeya_ Dzivakova; **19:** NPavelN, Oleksandr Khoma; **20:** KanKhem, Nadzeya_Dzivakova; **21:** Tetiana Lazunova, Tigatelu; **22:** feirin; **23:** MicrovOne, adekvat; **24:** MicrovOne, Armation74; **25:** marina_ua, AntonioFrancois; **26:** saenal78, Faisal Farabi; **27:** StudioBarcelona, lattesmile; **28:** Chuhail, Ilyabolotov, Abscent84, feirin; **29:** rubynurbaidi; **31:** marina_uapleshko74, Shendart; **33:** Tigatelu; **35:** aaaniram; **37:** AzFree; **39:** Diane Labombarbe, -ELIKA-, Victor_85, Photoplotnikov, RedlineVector, NPavelN; **41:** saenal78, LeeYenz, eliflamra; **47:** LeeYenz; **50:** VectorMoon, lunar_cat, ONYXprj, Abscent84, vectorsmarket, ChubarovY, drogatnev, Inkant, Denis08131; **51:** mejnak, luplupme, Angyee054, ONYXprj, Brillianata, Tetkoren, m_nesvit, Aleksangel, Artis777; **63:** LeeYenz; **64:** LeeYenz, saenal78; **70:** Maike Hildebrandt; saenal; **71** : adekvat; **72:** MicrovOne, feirin, ONYXprj, Maike Hildebrandt, Tetiana Lazunova; **73:** Tigatelu; **76:** Faisal Farabi; **81:** dashadima, eliflamra; **82:** colematt; **83:** colematt; **84:** colematt; **85:** colematt; **86:** colematt, **87:** colematt; **88:** Brillianata; Tigatelu; **89:** Faisal Farabi, adekvat, LeeYenz; **90:** colematt; **91:** terdpong pangwong; **92:** colematt; **93:** terdpong pangwong, Dar_ria; **94:** colematt; **95:** Tomacco; **96-97:** Tomacco, sabelskaya, MicrovOne,blueringmedia, Tetiana Lazunova, mejnak, MatoomMi, chekat; **98-99:** Maike Hildebrandt, Irina Griskova; **100:** AzFree, VladSt; **101:** Irina Griskova; **102:** blueringmedia, colematt; **103:** MicrovOne, bestsale, feirin, VectorMoon, luplupme, maryliflower, Brillianata, Irina Griskova; **104:** StudioBarcelona; **105:** Betsart; **106:** Eloku, Octopus 182; **107**: luplupme, VectorMoon, feirin, Maike Hildebrandt, Irina Griskova, saenal78, vectorsmarket, ksuklein, mart_m; **108**: vectorsmarket, Irina Griskova, m_nesvit, lattesmile, heriondesigncom; **109:** Betsart; **110:** Brillianata, ONYXprj; **111:** Betsart; **112:** colematt.

Preface

Collins Little Learners, comprising 20 books in all, provides graded, age-appropriate learning opportunities targeting holistic skill development for children in the age group of 3 years to 6 years. Spread across three levels, the series spans **English**, **Maths**, **General Awareness**, **Art and Craft** and **Hindi**. The series comes bundled with a **workbook** for English and a book of **rhymes and stories** for each level. The books use a combination of hands-on activities, frequent recapitulation and child-friendly artwork to facilitate learning.

The early years are the most significant years for growth, development and learning. During this period, children need adequate opportunities for stimulating experiences as these favourably affect various learning abilities. Learning experiences, when chosen and planned properly, fuel the innate curiosity of children, and stimulate their thought processes.

Learning and Development During the Early Years

The goal of learning during the early years is to enhance the all-round development of personality and to inculcate healthy attitudes, life skills and good values. Creative play, well thought-out games and developmentally appropriate activities help children develop the necessary skills to achieve these goals. The skills that children need to acquire can broadly be categorized as:

Motor: Motor skills pertain not only to gross motor activities like walking and jumping but also include fine motor activities like the ability to hold a pencil and write. The development of motor skills is facilitated through several graded activities.

Personal, Social and Emotional: Personal, social and emotional development, necessary for the overall personality development of children, is achieved using a combination of hands-on activities, role plays, and thought-provoking stories and rhymes.

Linguistic: The ability to listen, speak, read and write is critical to success in formal learning environments.

Mathematical: The development of mathematical ideas and concepts is important at this stage since children will continue to build on this learning for most of their years of schooling.

Knowledge and Understanding of the World: This is achieved through activities that enable children to explore their surroundings at home and in the school, and interact with the people that they see around them.

Health and Safety: The idea of health and safety is introduced through activities, stories and retelling of real-life experiences to encourage children to focus on health and well-being.

Creative: Creative skills are developed through a host of activities involving art, craft, painting, roleplays, etc.

Logical and Rational Thinking: Activities that involve simple puzzles, pattern-completion and visual discrimination help children to develop logical and rational thinking skills.

The Collins Little Learners Teacher Support Materials

The series includes extensive teacher support in the form of:

Teacher's Resource Books: Comprises stories, activity banks, classroom management and assessment guidelines

Multimedia eBook: Consists of animations, audio, video, slide shows, and interactive exercises

Print-ready Worksheets: Additional worksheets for practice and reinforcement, present in the digital support

Digital Flashcards: Flashcards with embedded sounds and animations for letter formation and letter-sound association

Animations and Songs: Animated stories and rhymes set to tune

Interactive Games: Interactivities to reinforce concepts in a fun and engaging manner

A note on the activity banks (included in the Teacher's Manuals)

The activity banks comprise hands-on activities that will help children develop the requisite skills to attain age-appropriate learning goals. In keeping with Howard Gardener's theory of multiple intelligences, the activity banks include options and variations that enable teachers to provide a planned cycle of experiences and opportunities, which foster each type of intelligence, to every child in their classrooms. The activity banks equip teachers to use the strengths of children as a means of complementing their less developed areas by valuing and nurturing the inherent abilities of the children.

Contents

Walkthrough

New concepts introduced through short rhymes, followed by practice pages for reinforcement

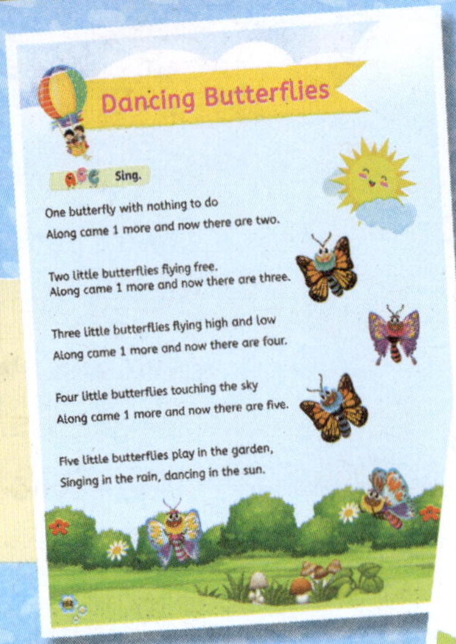

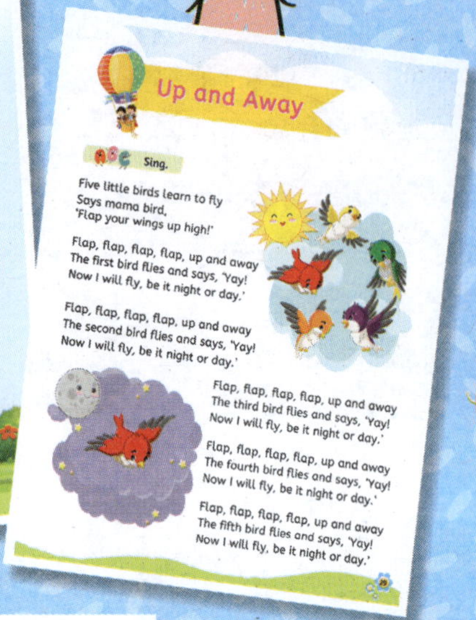

Counting beyond 10 continued using groups of 10

Number formation with pattern tracing

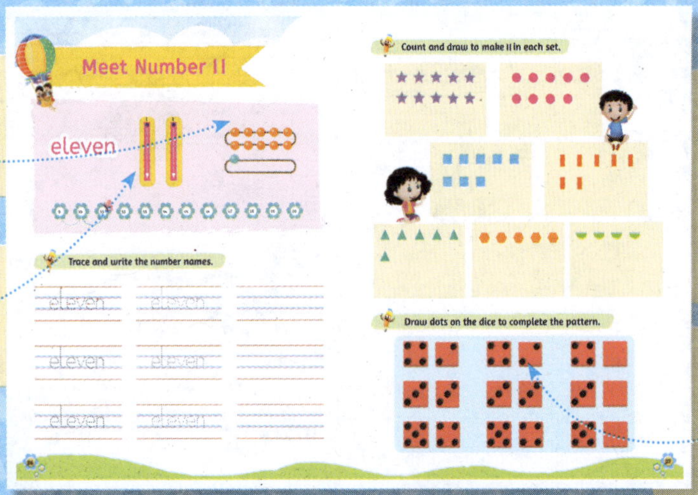

Exercises to lay the foundation for number operations, which are introduced at the end of the book

Exercises in interesting formats to develop observation and logical thinking skills

Number bonds introduced as a precursor to the concepts of addition and subtraction

Frequent revision pages in various formats for recapitulation and reinforcement

Multiple opportunities for revision of rote counting

I Know Numbers from 1 to 20

Match the pictures to the correct numbers.

1

19

11

9

3

Colour the correct number of objects.

15

7

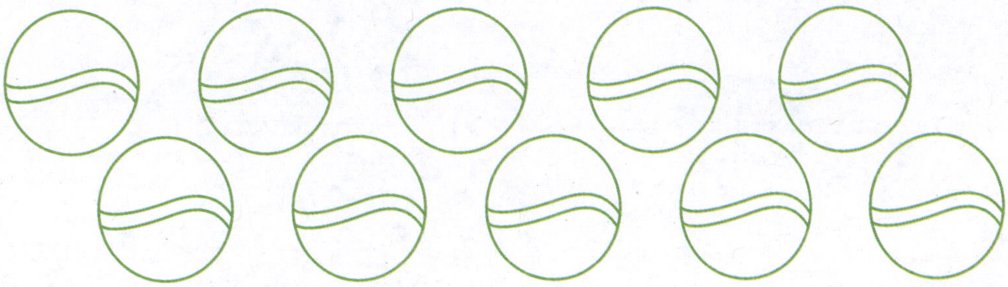

4

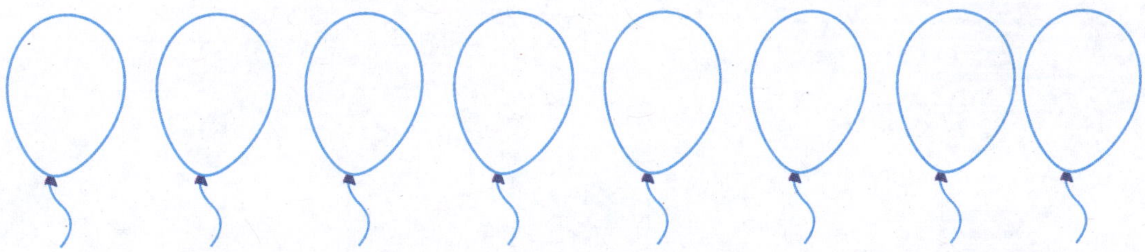

Skip count by 2's. Colour the correct boxes in red. One has been done for you.

1	2	3	4	5
6	7	8	9	10
11	12	13	14	15
16	17	18	19	20

I Know Counting from 1 to 50

Trace and write.

1	11	21	31	41
2	12	22	32	42
3	13	23	33	43
4	14	24	34	44
5	15	25	35	45
6	16	26	36	46
7	17	27	37	47
8	18	28	38	48
9	19	29	39	49
10	20	30	40	50

Colour the number that comes before.

43 44

22 23

13 14

Cross out (✗) the number that comes after.

9 10

44 45

29 30

Circle the number that comes in between.

9 10 11

30 31 32

42 43 44

I Can Count by Making Groups of 10

Colour the beads to make the given number.

11 and

12 and

13 and

14 and

15 and

16 and

17 and

18 and

19 and

20 and

I Know!

Draw a big box.

small

big

Colour the thick pencil.

Cross out (✗) the short ladder.

Draw a fat snowman.

Draw an apple near the boy.

Circle the different object.

Cross out (✗) the empty vase.

Colour the animal with the long neck.

Colour the light object.

Colour the rats going up the wall.

Draw fruits to fill the basket. Colour the picture.

Circle the same animals.

Tall, Taller, Tallest

Point and read.

tall taller tallest

Tick (✓) the tallest building.

Draw a taller lamppost here.

Write 1 for the tallest ladder, 2 for the taller ladder and 3 for the tall ladder.

Short, Shorter, Shortest

Point and read.

short shorter shortest

Tick (✓) the shortest cake.

Draw a short bat here.

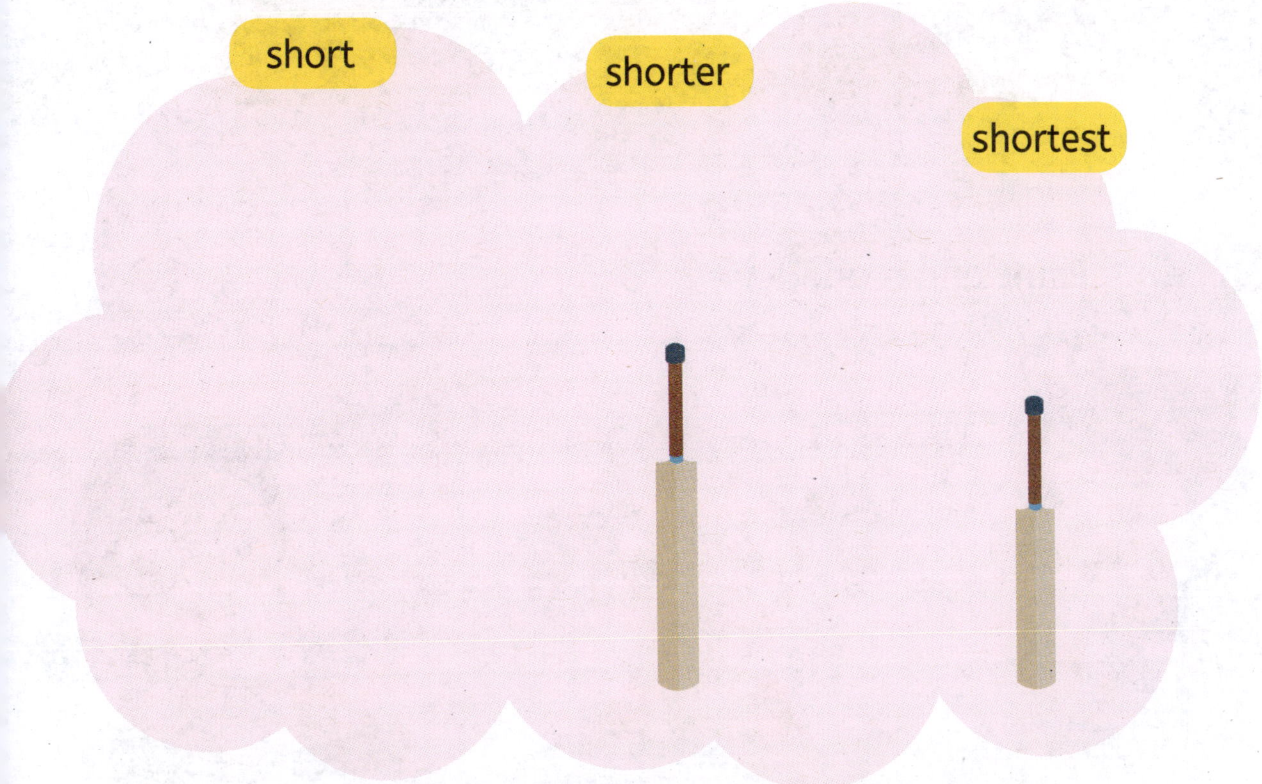

Write 1 for the shortest table, 2 for the shorter table and 3 for the short table.

Heavy, Heavier, Heaviest

Point and read.

heavy heavier heaviest

Tick (✓) the heaviest fruit.

☐ ☐ ☐

Colour the heaviest ●–grey, heavier ●–blue and heavy ●–black.

Write 1 for the heaviest lamb, 2 for the heavier lamb and 3 for the heavy lamb.

Thin, Thinner, Thinnest

Point and read.

Tick (✓) the thinnest candle.

Draw a thinner candy here.

thin

thinner

thinnest

Write 1 for the thinnest book, 2 for the thinner book and 3 for the thin book.

Big, Bigger, Biggest

Point and read.

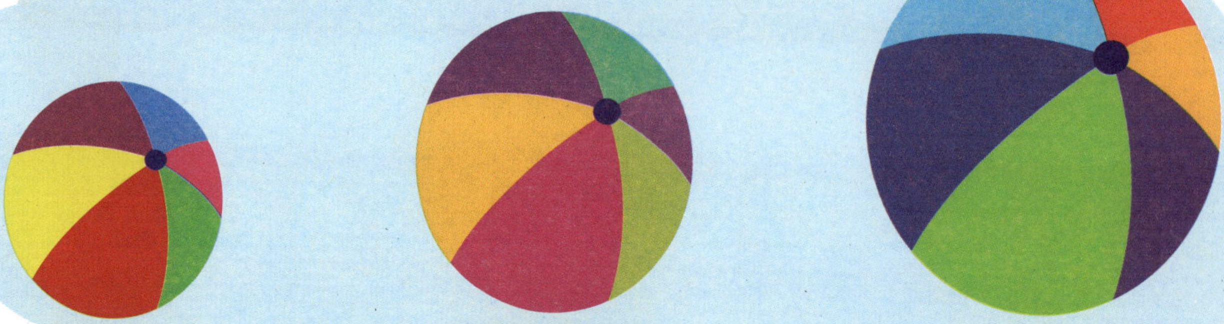

big bigger biggest

Tick (✓) the biggest bear.

Colour the biggest glass ●-red, bigger glass ●-blue and big glass ●-orange.

Write 1 for the biggest apple, 2 for the bigger apple and 3 for the big apple.

Small, Smaller, Smallest

Point and read.

Tick (✓) the smallest circle.

Draw a smaller flower here.

Write 1 for the smallest cap, 2 for the smaller cap and 3 for the small cap.

I Know!

Circle the heavier pumpkin.

Circle the shortest lamppost.

Colour the biggest bee.

Colour the smallest ball.

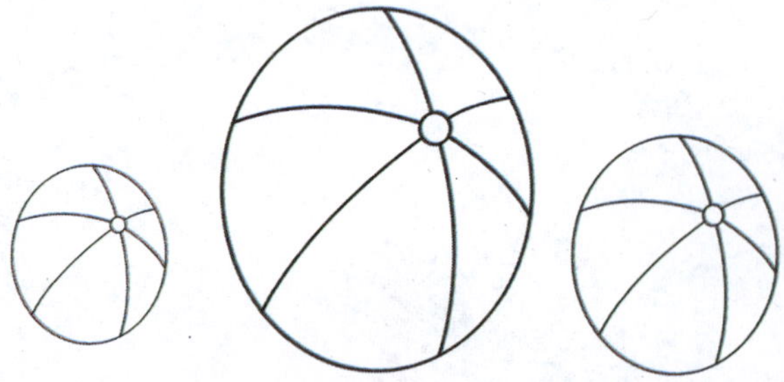

Tick (✓) the thinnest pencil.

Tick (✓) the tallest candle.

Up and Away

Sing.

Five little birds learn to fly
Says mama bird,
'Flap your wings up high!'

Flap, flap, flap, flap, up and away
The first bird flies and says, 'Yay!
Now I will fly, be it night or day.'

Flap, flap, flap, flap, up and away
The second bird flies and says, 'Yay!
Now I will fly, be it night or day.'

Flap, flap, flap, flap, up and away
The third bird flies and says, 'Yay!
Now I will fly, be it night or day.'

Flap, flap, flap, flap, up and away
The fourth bird flies and says, 'Yay!
Now I will fly, be it night or day.'

Flap, flap, flap, flap, up and away
The fifth bird flies and says, 'Yay!
Now I will fly, be it night or day.'

Ordinal Numbers

Look at the picture. Point and read.

first | second | third | fourth | fifth

Match the child to the position.

fifth | fourth | third | second | first

Circle the second fruit. Count from the left.

Colour the fifth glass. Count from the left.

Draw a cup near the third bear. Count from the left.

Cross out (✗) the fourth toy. Count from the left.

Number Names from 1 to 10

Trace and write.

1	one	one	one
2	two	two	two
3	three	three	three
4	four	four	four
5	five	five	five

Trace and write.

6	six	six	six
7	seven	seven	seven
8	eight	eight	eight
9	nine	nine	nine
10	ten	ten	ten

Match the number to the correct number name.

1	two
2	seven
3	nine
4	one
5	eight
6	six
7	three
8	five
9	ten
10	four

Look at the picture.

Count and write the correct number names.

   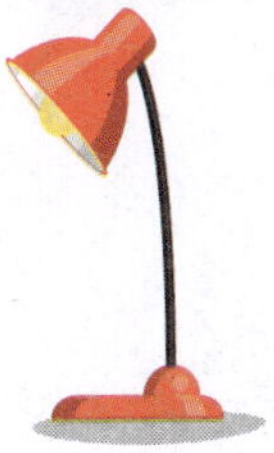 

one				three

Meet Number 11

eleven

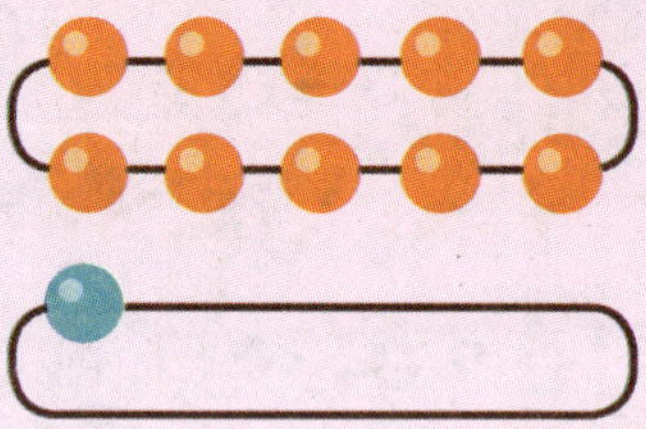

Trace and write the number names.

eleven eleven

eleven eleven

eleven eleven

Count and draw to make 11 in each set.

Draw dots on the dice to complete the pattern.

Meet Number 12

twelve

12

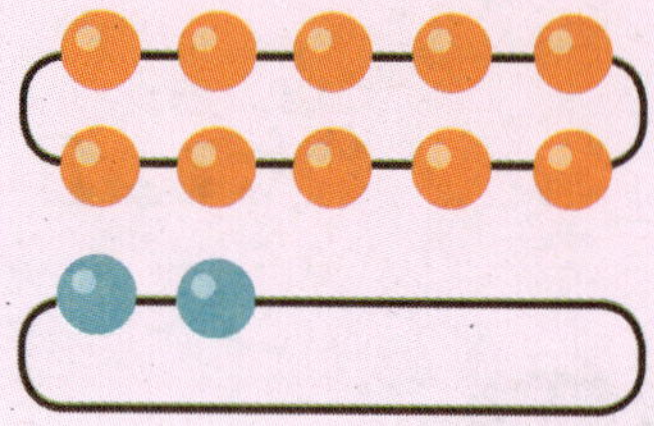

9 10 11 12 13 14 15 16 17 18 19 20

Trace and write the number names.

twelve twelve

twelve twelve

twelve twelve

Count and draw to make 12 in each set.

Match the things that go together.

Meet Number 13

thirteen

13

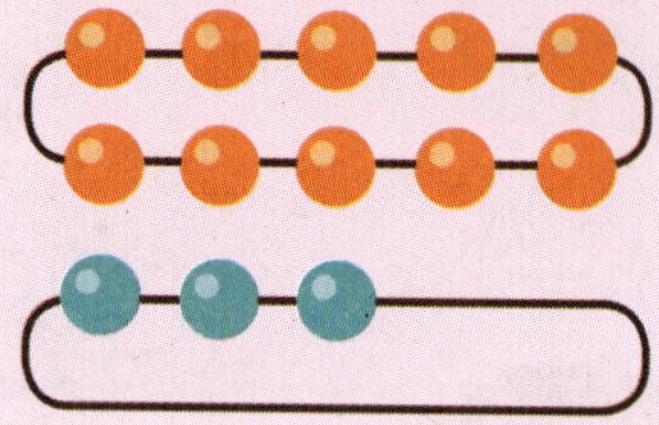

9 10 11 12 13 14 15 16 17 18 19 20

Trace and write the number names.

thirteen thirteen

thirteen thirteen

thirteen thirteen

Count and draw to make 13 in each set.

Match the things that go together.

Meet Number 14

fourteen

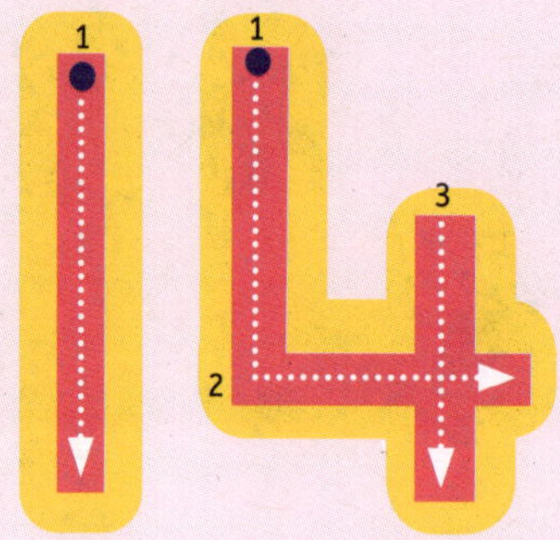

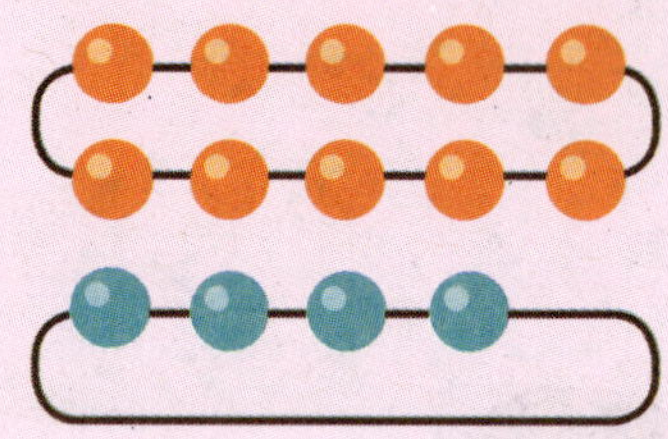

Trace and write the number names.

fourteen fourteen

fourteen fourteen

fourteen fourteen

Count and draw to make 14 in each set.

Meet Number 15

fifteen

15

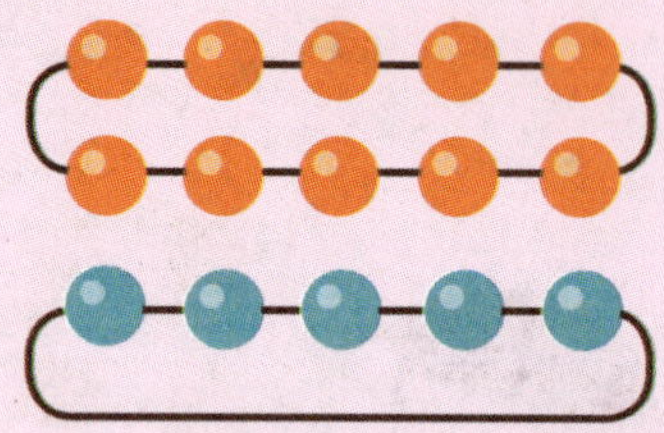

9 10 11 12 13 14 15 16 17 18 19 20

Trace and write the number names.

fifteen fifteen

fifteen fifteen

fifteen fifteen

Count and draw to make 15 in each set.

I Know Number Names from 11 to 15

Match the numbers to the number names.

11 12 13 14 15

thirteen eleven fourteen fifteen twelve

Fill in the missing letters.

e___ ___ven

thirt___ ___ ___

t___ ___ lv___

fift___ ___ ___

fourt___ ___ ___

Trace and write.

eleven

twelve

thirteen

fourteen

fifteen

Shapes

Point and read.

triangle

circle

square

rectangle

star

kite

semicircle

Trace and colour the circle.

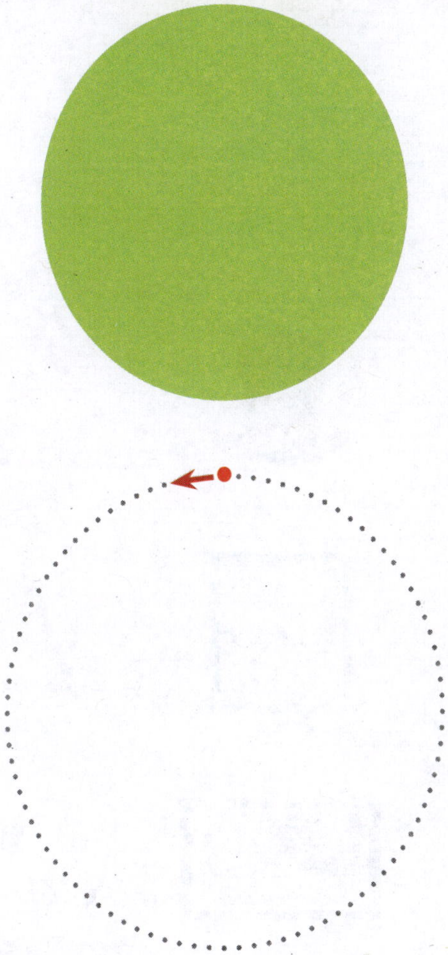

Trace and colour the star.

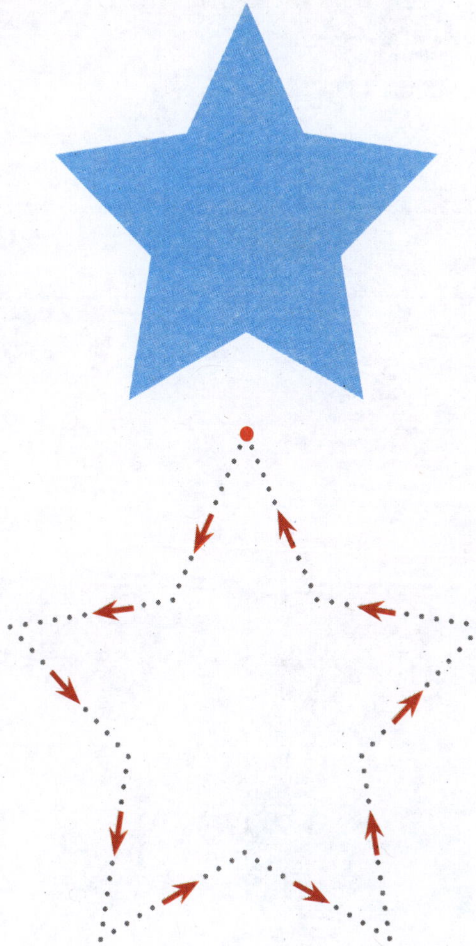

Trace and colour the semicircle.

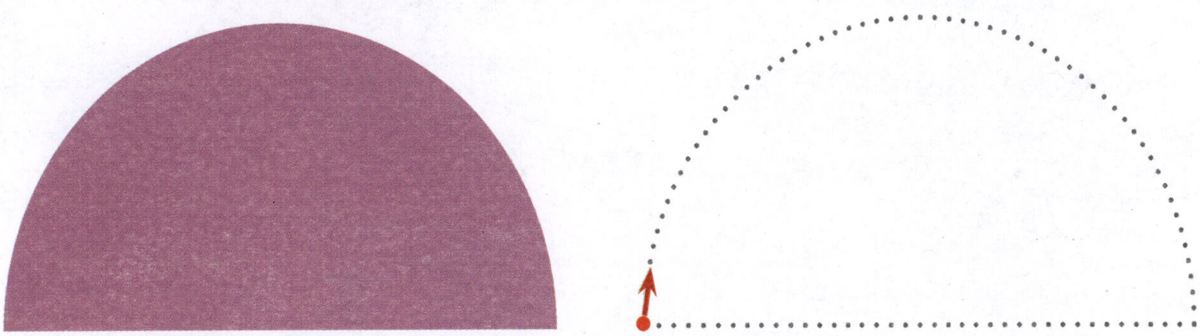

Shapes Around Us

Cross (✗) the odd one out.

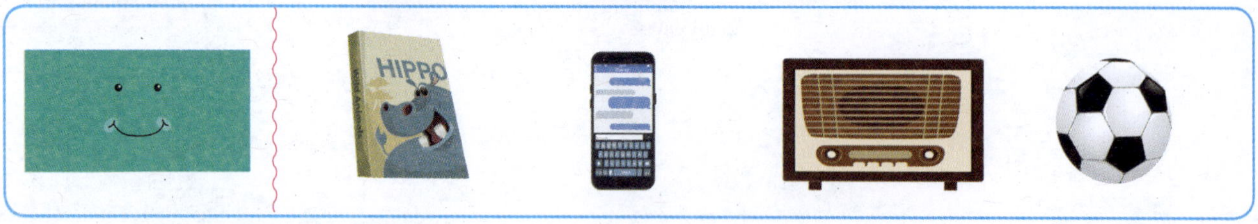

Draw a picture for each shape. Colour the pictures.

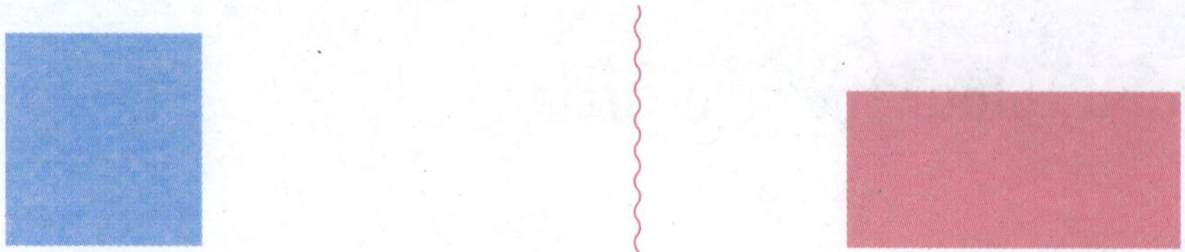

Cross (✗) the odd one out.

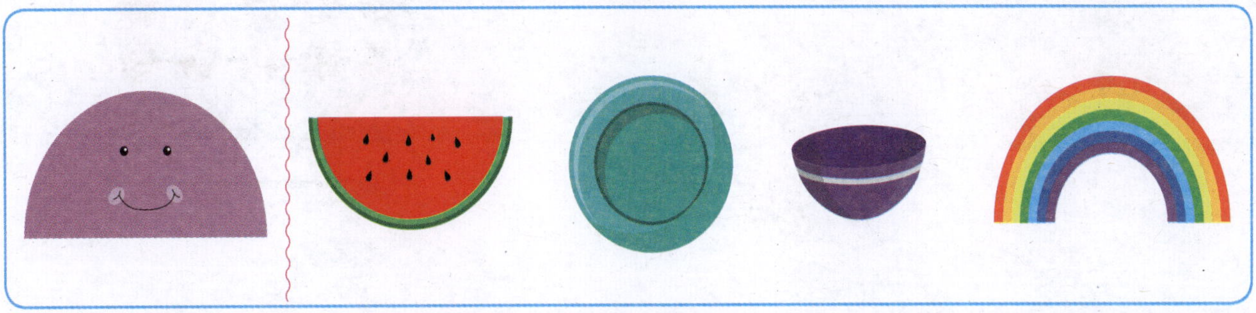

Draw a picture for each shape. Colour the pictures.

Meet Number 16

sixteen

16

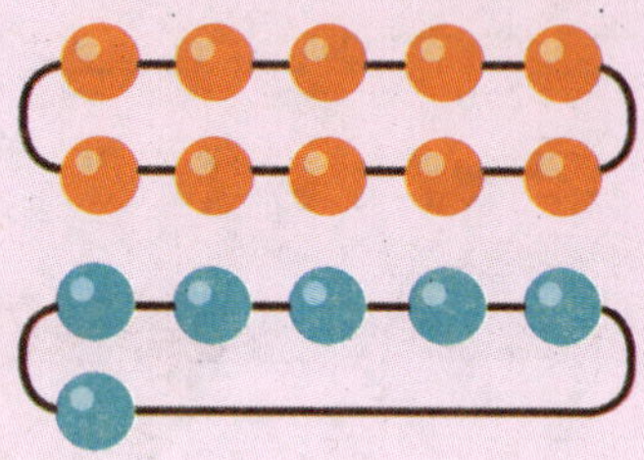

9 10 11 12 13 14 15 16 17 18 19 20

Trace and write the number names.

sixteen sixteen

sixteen sixteen

sixteen sixteen

Count and draw to make 16 in each set.

Meet Number 17

seventeen

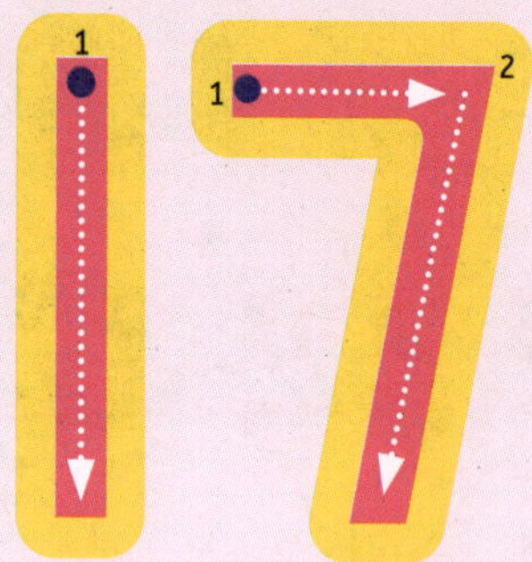

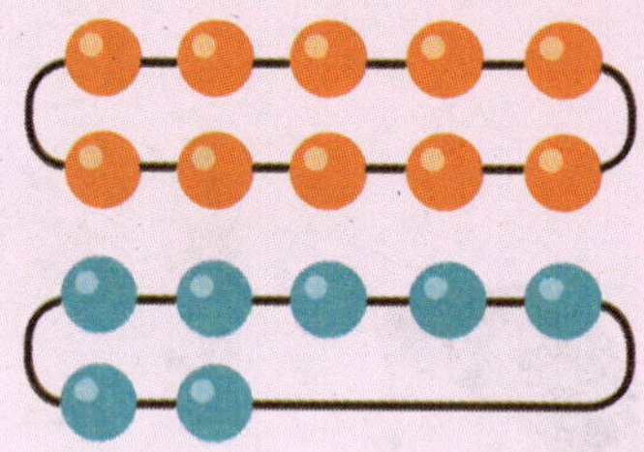

Trace and write the number names.

seventeen seventeen

seventeen seventeen

seventeen seventeen

Count and draw to make 17 in each set.

Meet Number 18

eighteen

18

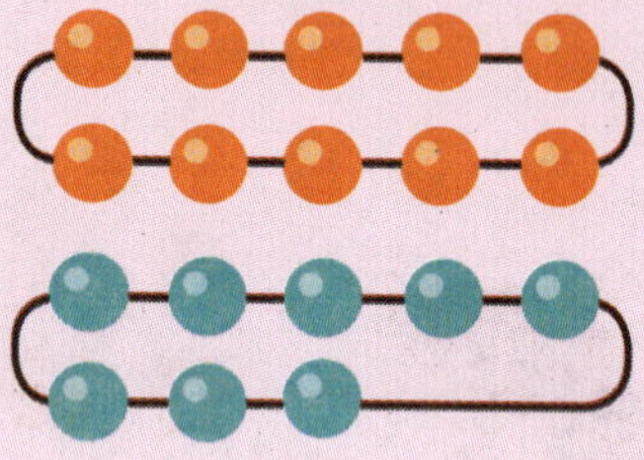

9 10 11 12 13 14 15 16 17 18 19 20

Trace and write the number names.

eighteen eighteen

eighteen eighteen

eighteen eighteen

Count and draw to make 18 in each set.

Meet Number 19

nineteen

19

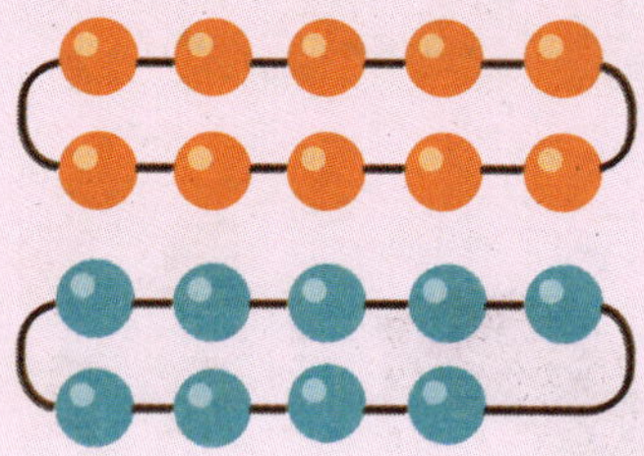

9 10 11 12 13 14 15 16 17 18 19 20

Trace and write the number names.

nineteen nineteen

nineteen nineteen

nineteen nineteen

Count and draw to make 19 in each set.

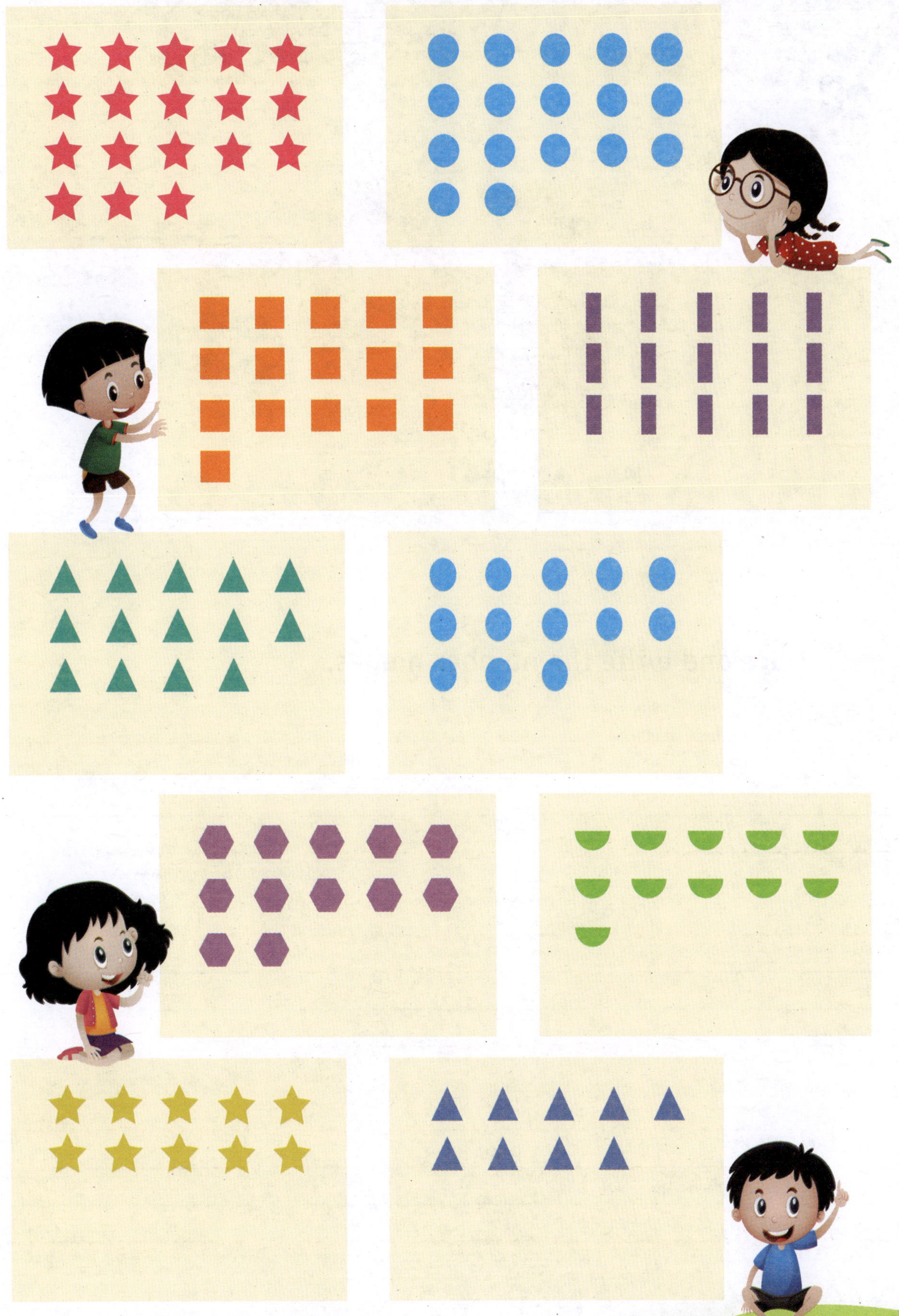

Meet Number 20

twenty

20

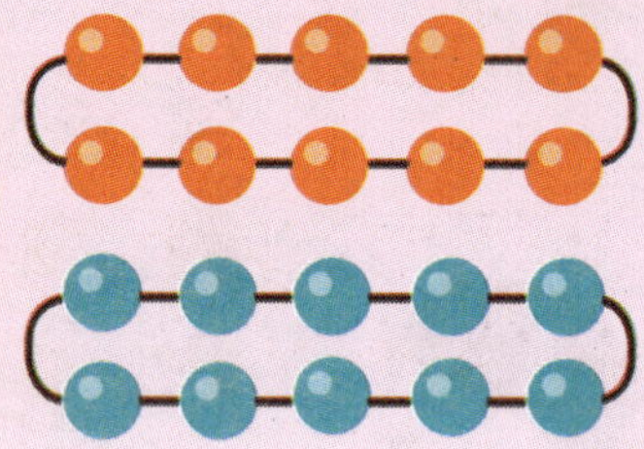

9 10 11 12 13 14 15 16 17 18 19 20

Trace and write the number names.

twenty twenty

twenty twenty

twenty twenty

Count and draw to make 20 in each set.

I Know Number Names from 16 to 20

Match the number names to the numbers.

nineteen   twenty   sixteen

seventeen   eighteen

16   17   18   19   20

Fill in the missing letters.

s___xt___en   s___v___nt___en   eight___ ___ ___

nin___tee___   tw___nty

Trace and write.

sixteen

seventeen

eighteen

nineteen

twenty

Revise Numbers 21 to 30

Read, trace and write.

21	22	23	24	25	26	27	28	29	30
21	22	23	24	25	26	27	28	29	30

Match the objects to the correct numbers.

25

30

23

Count and colour.

21 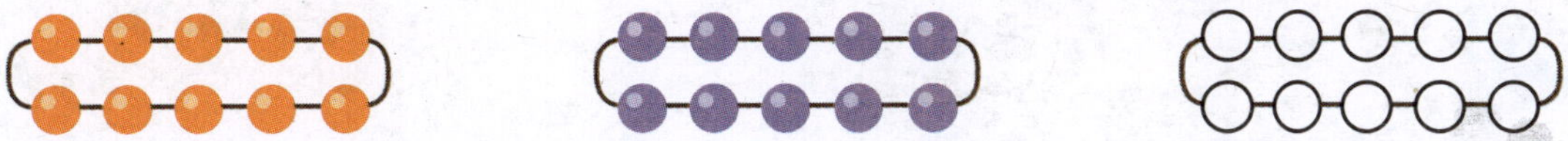

22 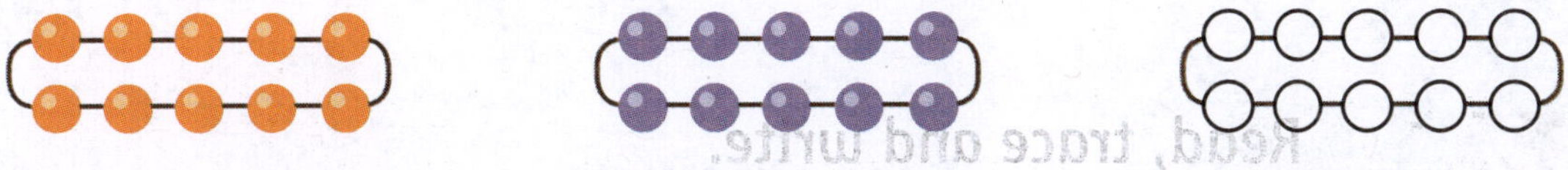

23 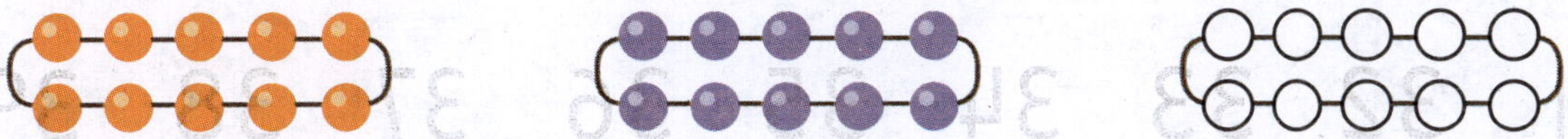

24 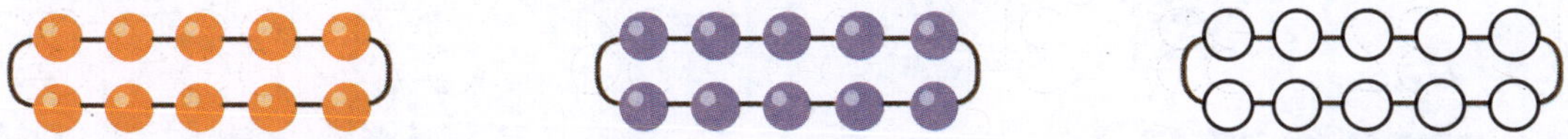

25 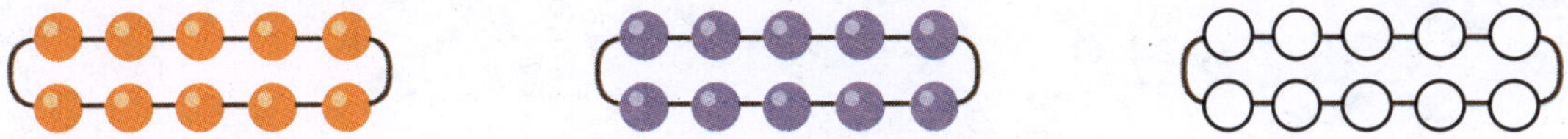

26 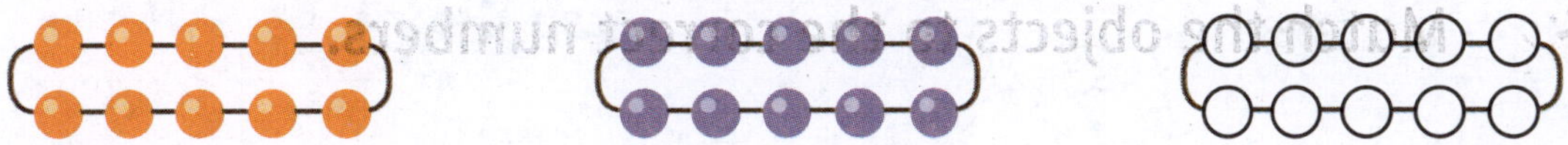

27 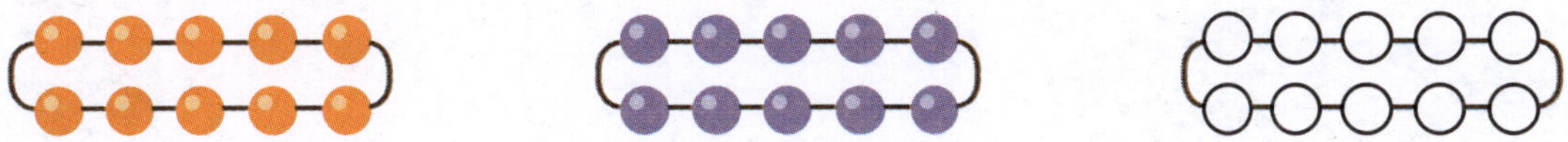

28 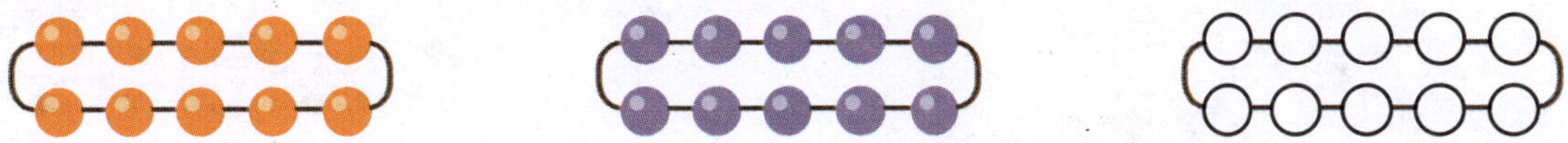

29 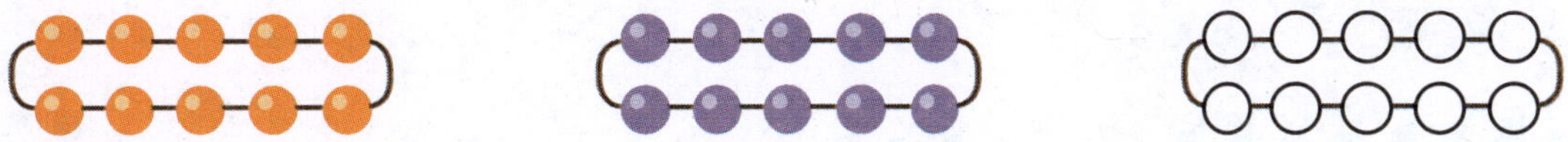

30 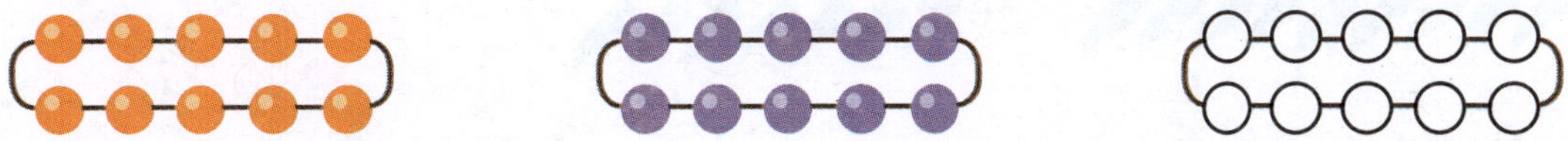

Revise Numbers 31 to 40

Read, trace and write.

31	32	33	34	35	36	37	38	39	40
31	32	33	34	35	36	37	38	39	40

Match the objects to the correct numbers.

 • • 35

 • • 37

 • • 40

Count and colour.

31 

32 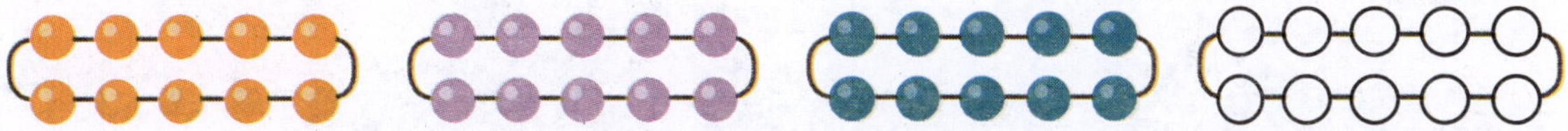

33 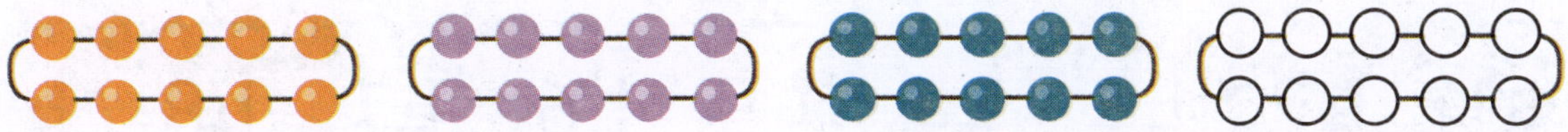

34 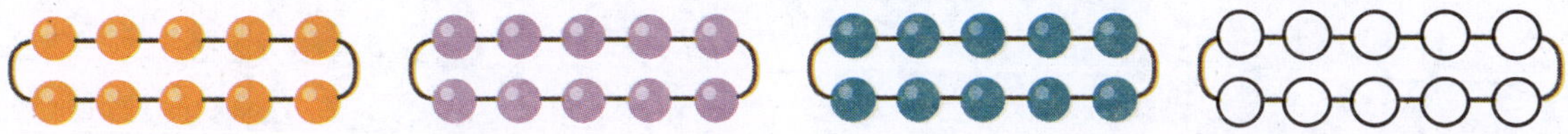

35 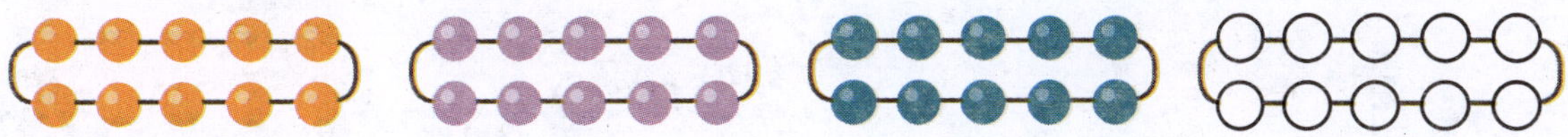

36 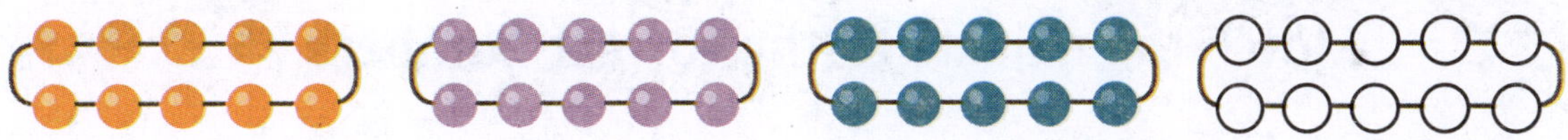

37 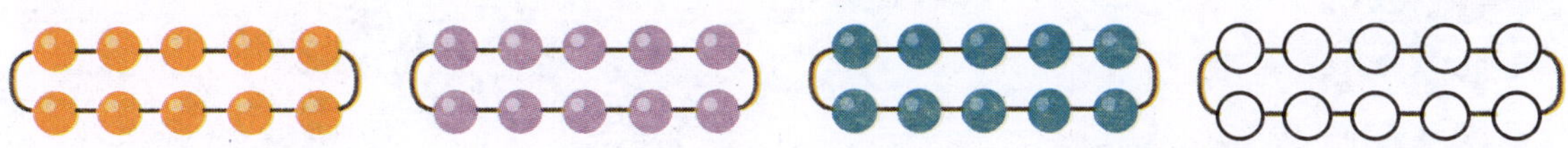

38 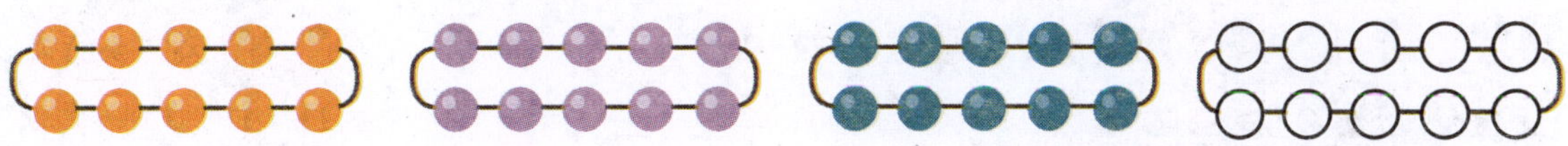

39 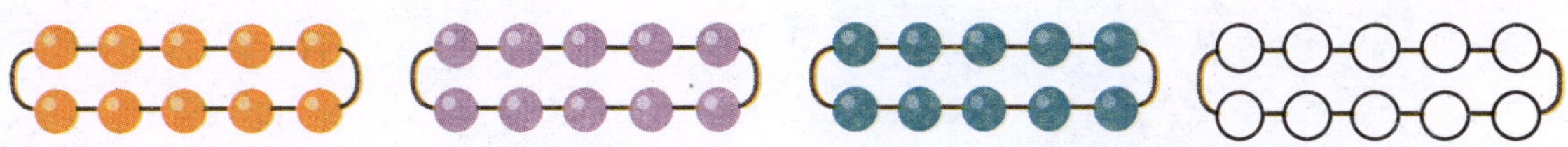

40 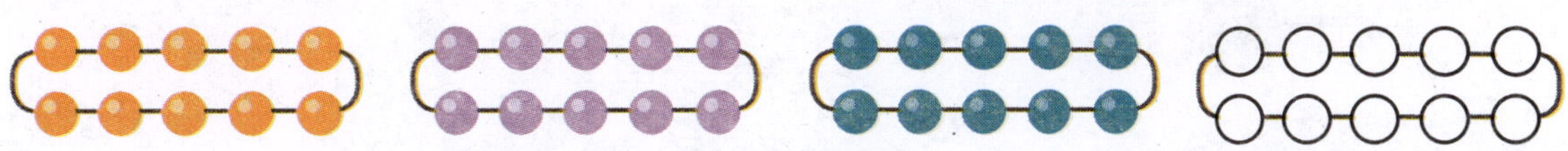

Revise Numbers 41 to 50

Read, trace and write.

41	42	43	44	45	46	47	48	49	50
41	42	43	44	45	46	47	48	49	50

Match the objects to the correct numbers.

46

42

50

Count and colour.

41 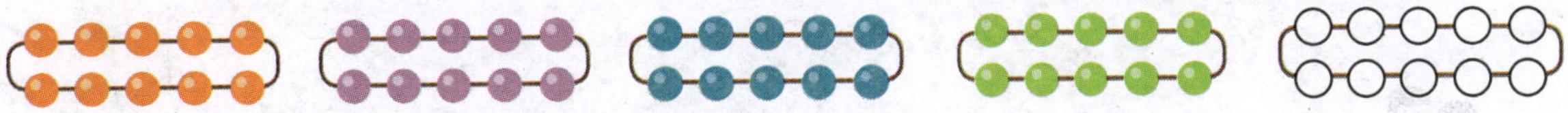

42 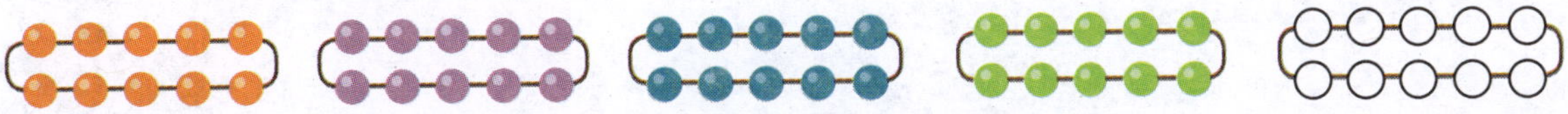

43 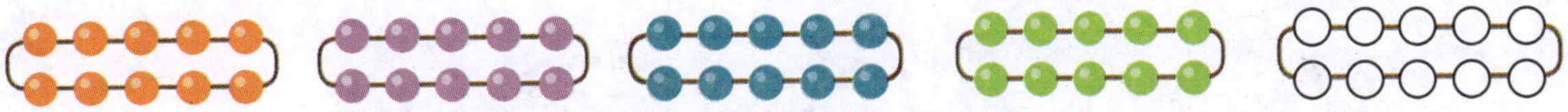

44 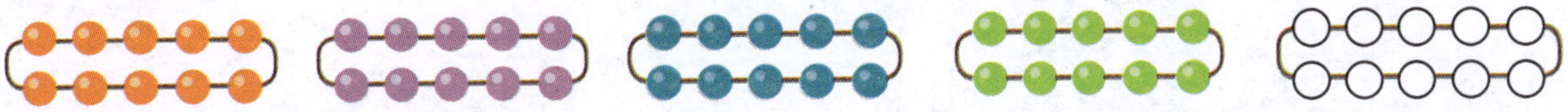

45 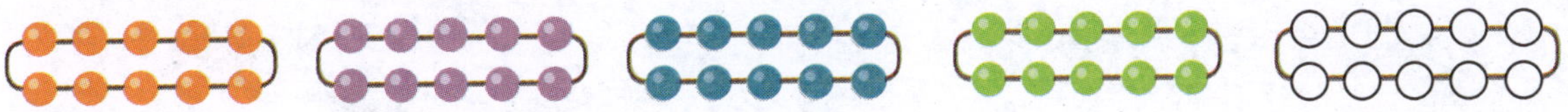

46 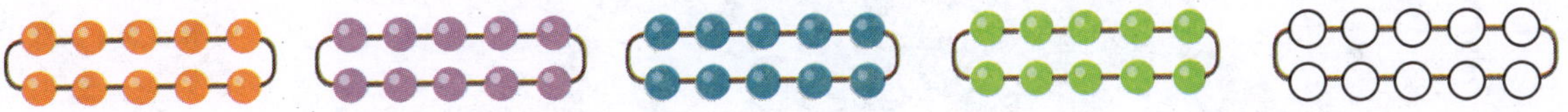

47 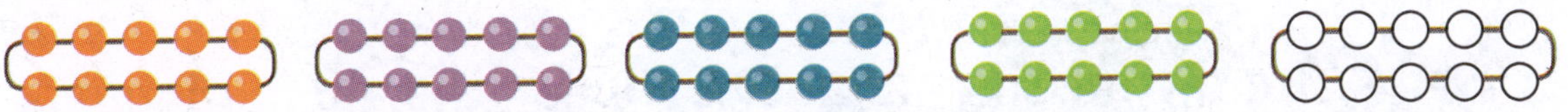

48 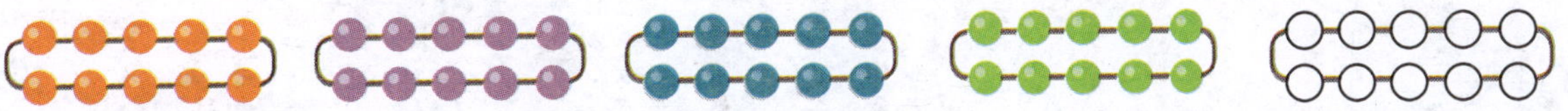

49 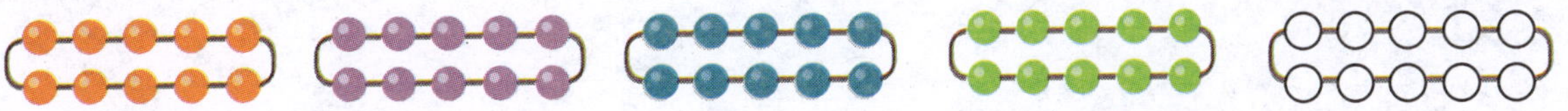

50 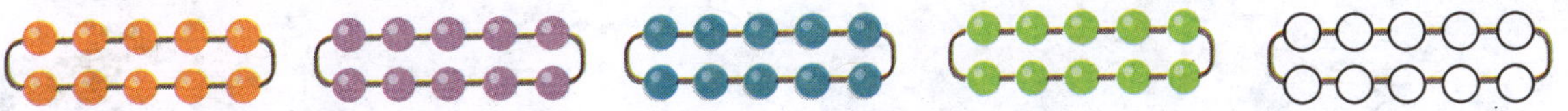

I Know Numbers Till 50!

Write the numbers that come before.

___ 50   ___ 49   ___ 27   ___ 42

Write the numbers that come after.

34 ___   42 ___   29 ___   37 ___

Write the missing number between the given numbers.

44 ___ 46   22 ___ 24

38 ___ 40   29 ___ 31

Numbers Till 50: More–Less

Colour the number that is more.

47 48 | 36 42 | 24 34

37 28 | 49 48 | 26 21

Cross out (✗) the one that is more.

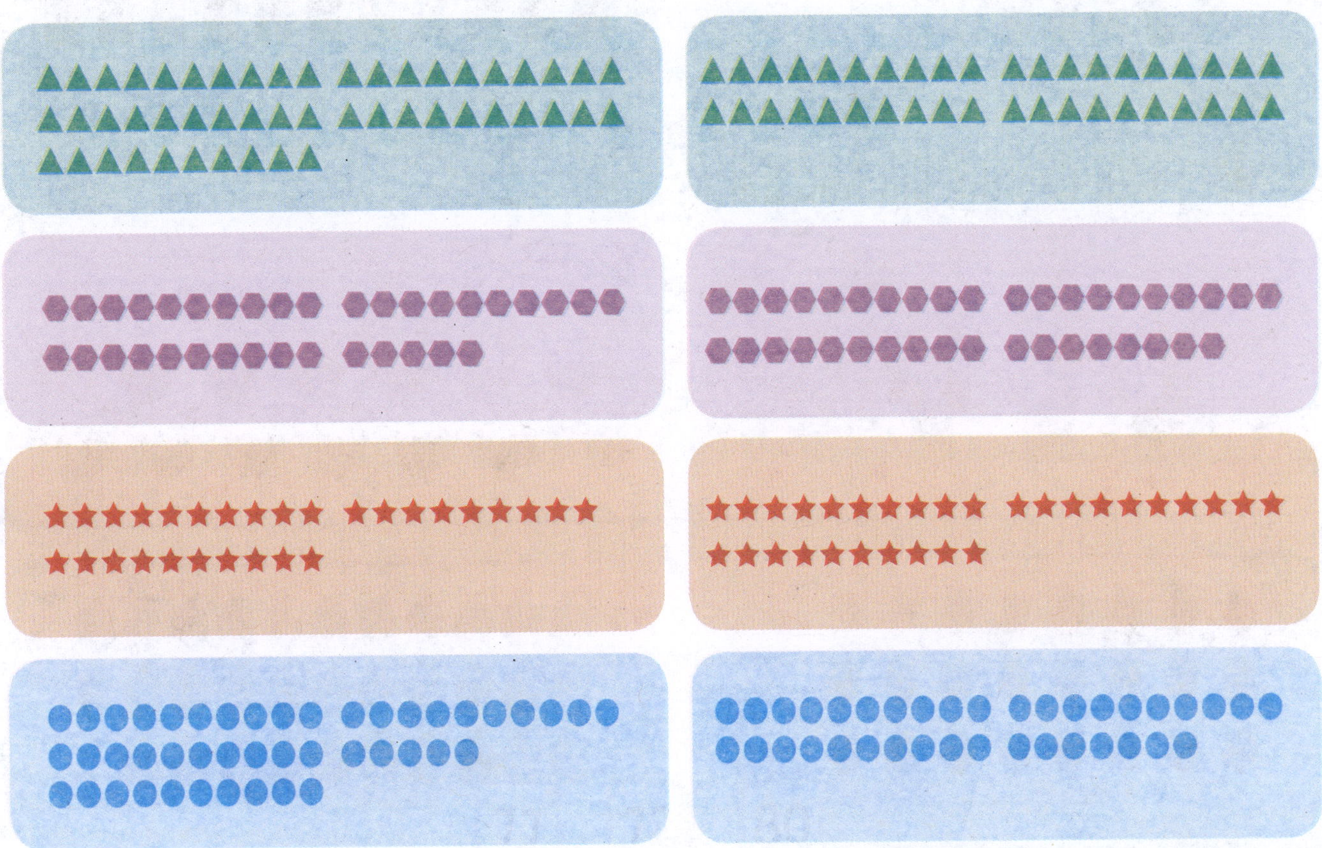

Big and Small Numbers

Cross out (✗) the small number in each set.

Tick (✓) the big number in each set.

Biggest and Smallest Numbers

Colour the smallest number blue. One has been done for you.

34 45 41

27 50 43

16 8 39

12 47 33

Colour the biggest number green and the smallest number yellow in each set.

10 20 30 40 50

12 20 25 32 44

17 19 31 43 50

I Can Arrange Numbers

Write the biggest number in the red box.
Write the smallest number in the blue box.

				Red box	Blue box
12	14	16	24		
18	22	25	30		
32	22	12	2		

Write the numbers from the biggest to the smallest.
One has been done for you.

				→				
34	11	45	23	→	45	34	23	11
22	29	14	50	→				
32	48	17	9	→				

Write the numbers from the smallest to the biggest.
One has been done for you.

30 12 44 50 → | 12 | 30 | 44 | 50 |

13 19 21 25 → | | | | |

50 2 34 21 → | | | | |

39 26 17 49 → | | | | |

19 28 37 46 → | | | | |

48 7 16 23 → | | | | |

Join the dots to complete the picture. Colour the picture.

Skip Counting by 2's

Look at the picture. Trace the path of the butterfly.

Colour every second number. One has been done for you.

1	2	3	4	5	6	7	8	9	10
11	12	13	14	15	16	17	18	19	20
21	22	23	24	25	26	27	28	29	30
31	32	33	34	35	36	37	38	39	40
41	42	43	44	45	46	47	48	49	50

Skip Counting by 3's

Look at the picture. Trace the path of the butterfly.

Colour every third number. One has been done for you.

1	2	3	4	5	6	7	8	9	10
11	12	13	14	15	16	17	18	19	20
21	22	23	24	25	26	27	28	29	30
31	32	33	34	35	36	37	38	39	40
41	42	43	44	45	46	47	48	49	50

Skip Counting by 5's

Look at the picture. Trace the path of the butterfly.

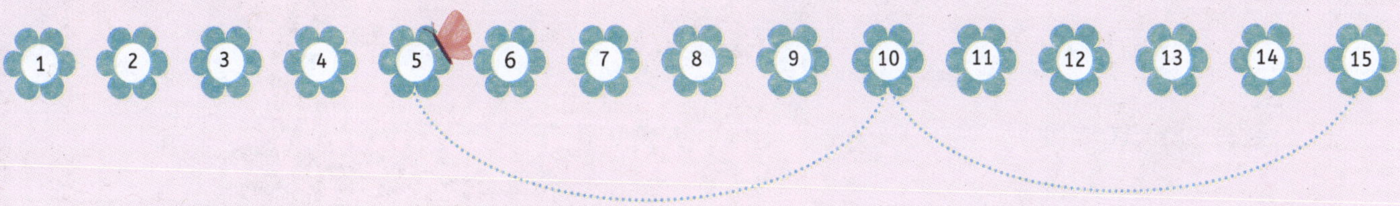

Colour every fifth number. One has been done for you.

1	2	3	4	5	6	7	8	9	10
11	12	13	14	15	16	17	18	19	20
21	22	23	24	25	26	27	28	29	30
31	32	33	34	35	36	37	38	39	40
41	42	43	44	45	46	47	48	49	50

Skip Counting by 10's

Look at the picture. Trace the path of the butterfly.

Colour every tenth number. One has been done for you.

1	2	3	4	5	6	7	8	9	10
11	12	13	14	15	16	17	18	19	20
21	22	23	24	25	26	27	28	29	30
31	32	33	34	35	36	37	38	39	40
41	42	43	44	45	46	47	48	49	50

Backward Counting

Count backwards from 50 to 1. Fill in the missing numbers.

50	40	30	20	10
41	31	21	11	1

Counting from 51 to 60

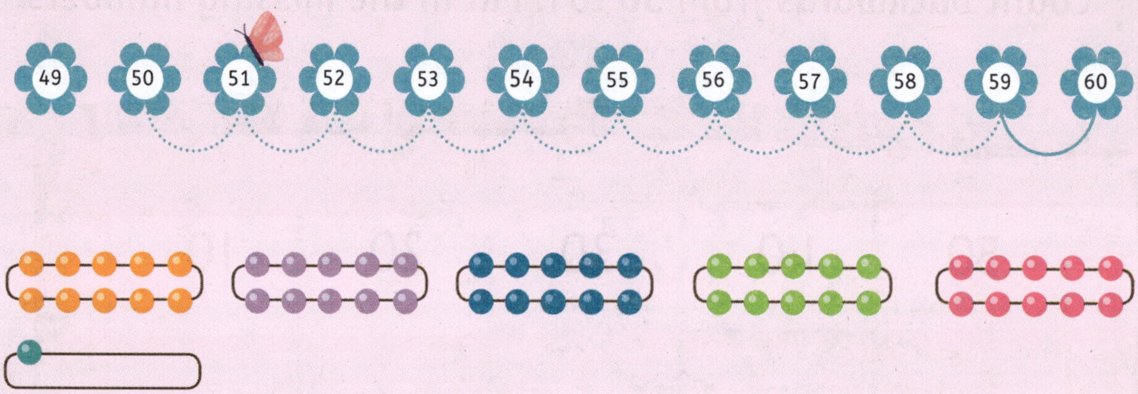

Trace and write.

51	51		56	56	
52	52		57	57	
53	53		58	58	
54	54		59	59	
55	55		60	60	

Count and tick (✔) the correct number.

55 54 58

56 51 60

53 60 58

Counting from 61 to 70

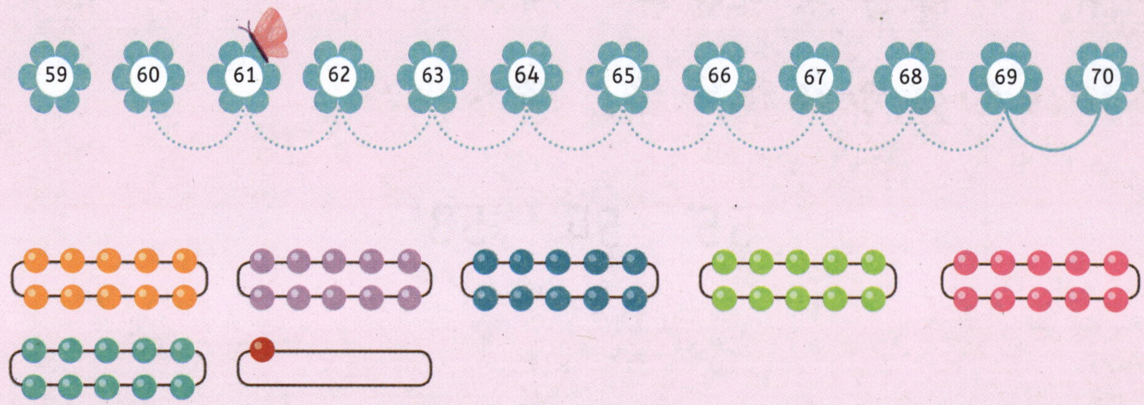

Trace and write.

61	61		66	66	
62	62		67	67	
63	63		68	68	
64	64		69	69	
65	65		70	70	

Count and tick (✓) the correct number.

68 64 61

66 61 65

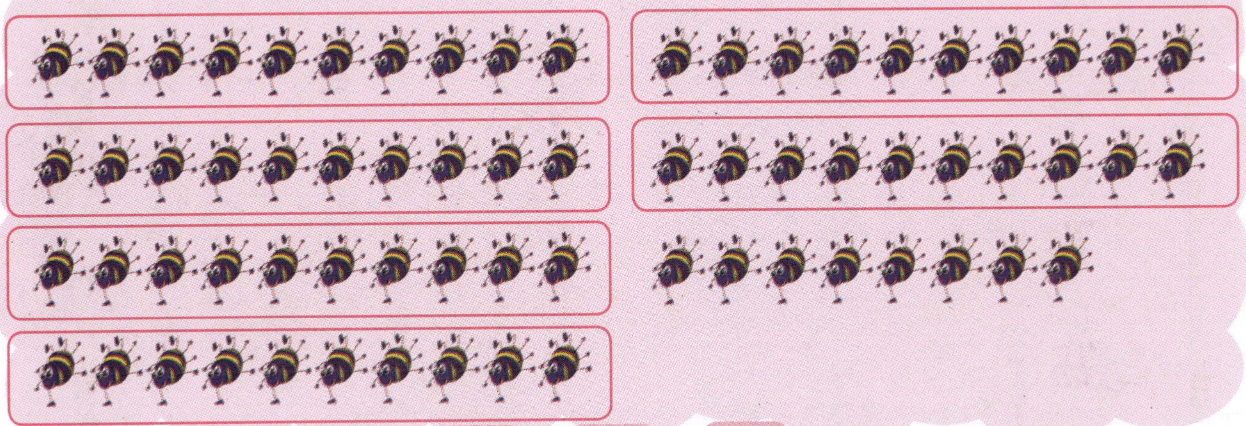

67 60 68

Counting from 71 to 80

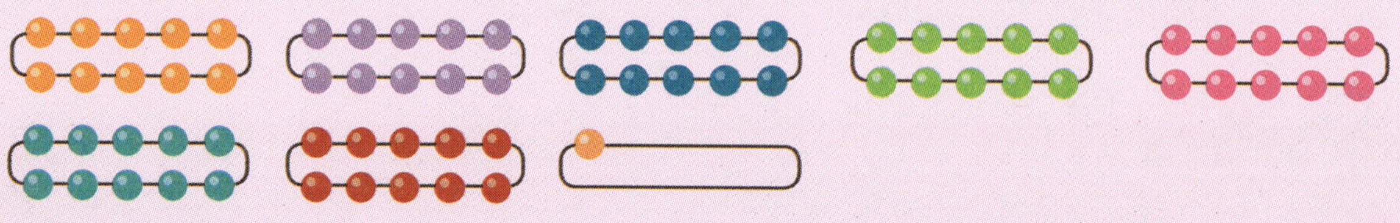

Trace and write.

71	71		76	76	
72	72		77	77	
73	73		78	78	
74	74		79	79	
75	75		80	80	

Count and tick (✔) the correct number.

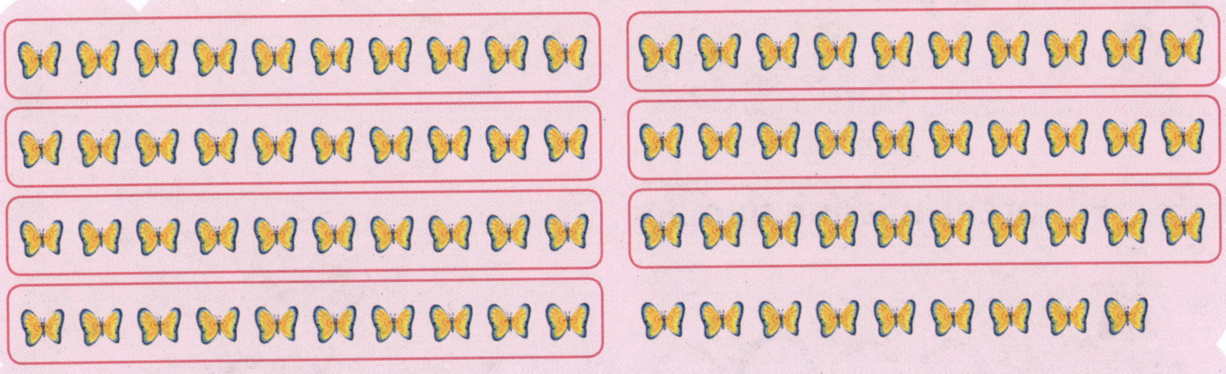

71 74 79

74 78 75

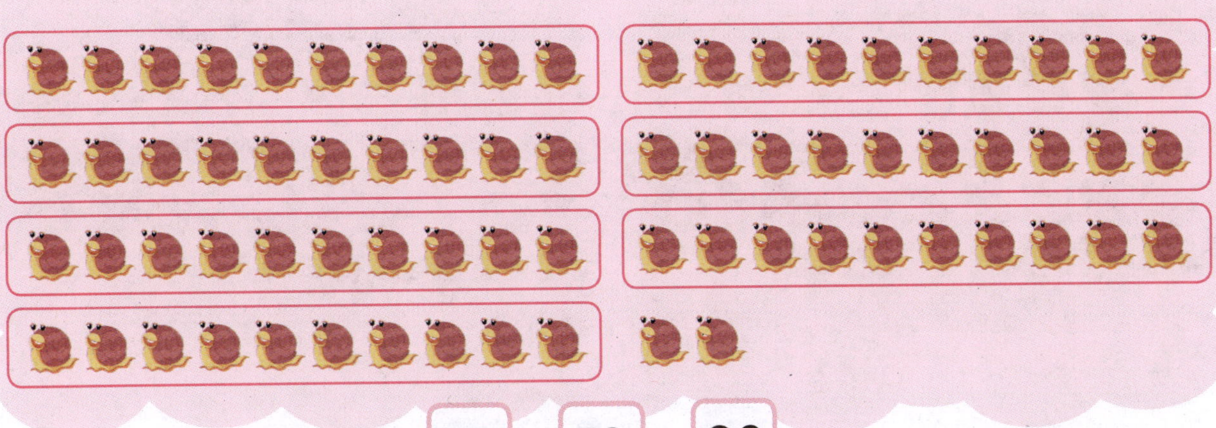

77 72 80

I Can Count Till 80!

Fill in the missing number.

51 52 ___ 54 ___ 56 57 58 ___ ___

61 62 ___ 64 ___ 66 67 ___ 69 ___

___ 72 73 ___ 75 76 ___ 78 79 ___

Tick (✔) the set with more objects.

☐

☐

Count and write.

Count and write.

Counting from 81 to 90

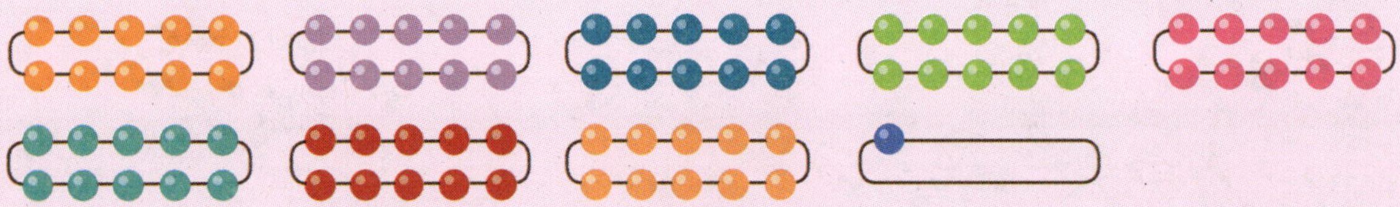

Trace and write.

81	81		86	86	
82	82		87	87	
83	83		88	88	
84	84		89	89	
85	85		90	90	

Count and tick (✓) the correct number.

82 86 90

83 90 88

82 85 88

Counting from 91 to 100

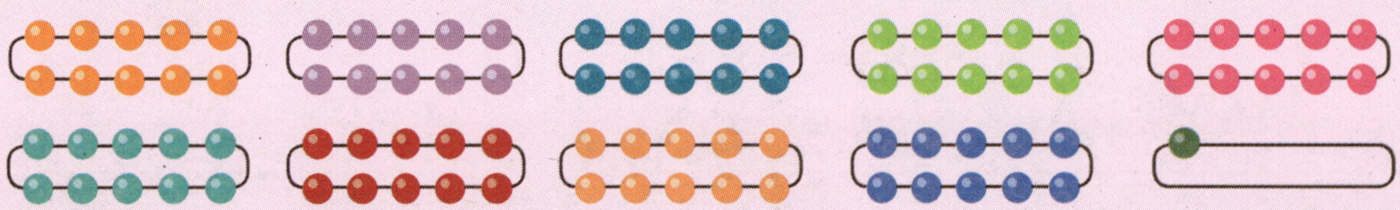

Trace and write.

91	91		96	96	
92	92		97	97	
93	93		98	98	
94	94		99	99	
95	95		100	100	

Count and tick (✔) the correct number.

94 96 99

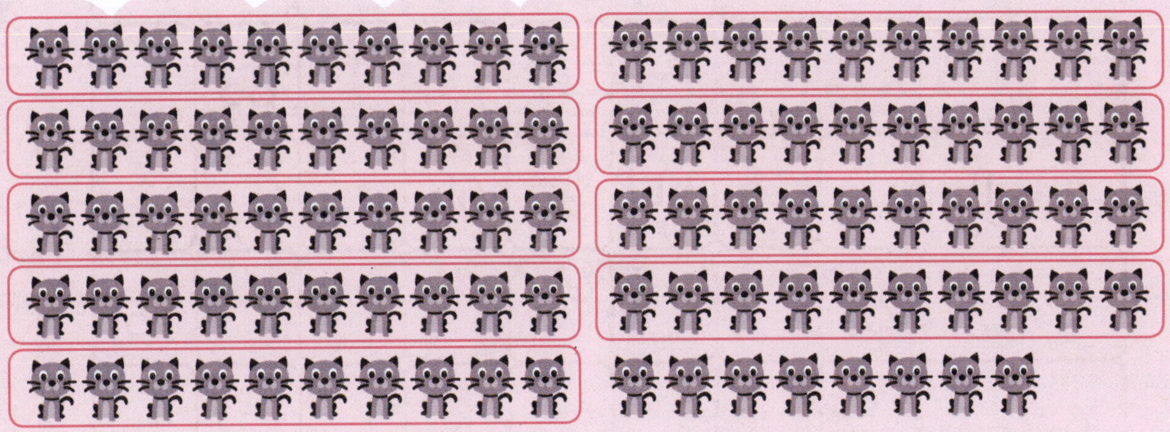

93 91 98

96 98 95

I Know Numbers from 1 to 100

Fill in the missing numbers.

1	11		31				71		
					52				
								83	
	14			44		64			
		25							95
6									
			37			67	77		
					58			88	
	19			49					
		30							100

Join the dots and colour.

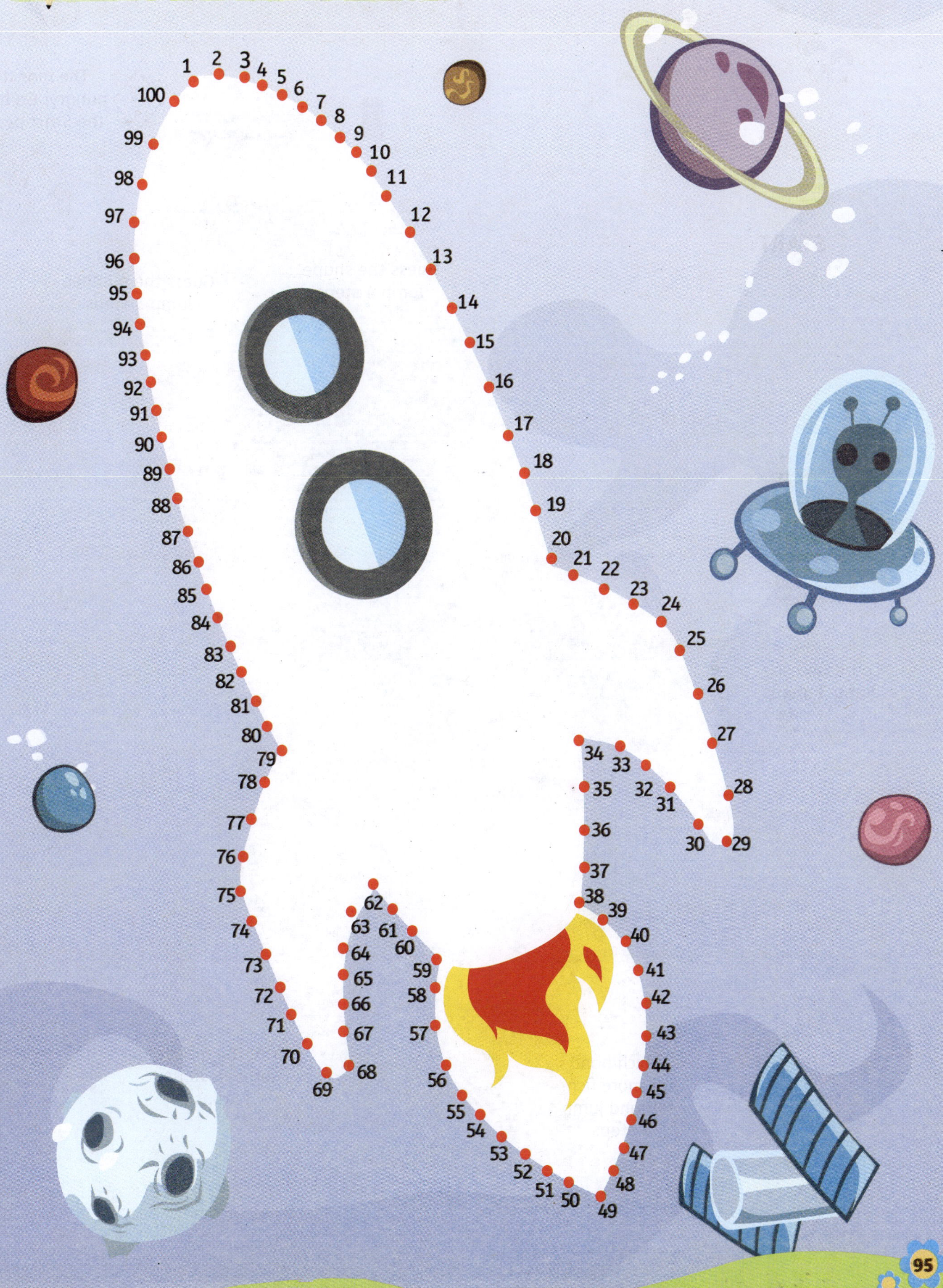

I Know Numbers from 1 to 100

rite number name. Jump 6 steps.
ele_ _ _
12 13 14 16 17 18 19 20
Tell the missing number. Jump 5 steps.
You are now taller than last year. Jump 2 steps.
tw_ _ty
Write number name. Jump 6 steps.
Guess the number. Jump 3 steps.
29 28 27 26 25 24 23 22 21
Pick the smallest kite. Jump 3 steps.
52 53 54 55 56 57 58 59
You had 3 cats. One ran away. Go back 2 steps.
Big, heavy elephant blocked the way. Go back to start.
68 67 66 65 64 63 62 61 60
92 93 94 95 96 97 98 99 100
Count and write the number. Jump 4 steps.
and
FINISH

Number Bonds

Count and write the number of blue dots.

10
5

10
6

10
7

10
8

10
9

Number Bonds

Draw the correct number of red dots.
One has been done for you.

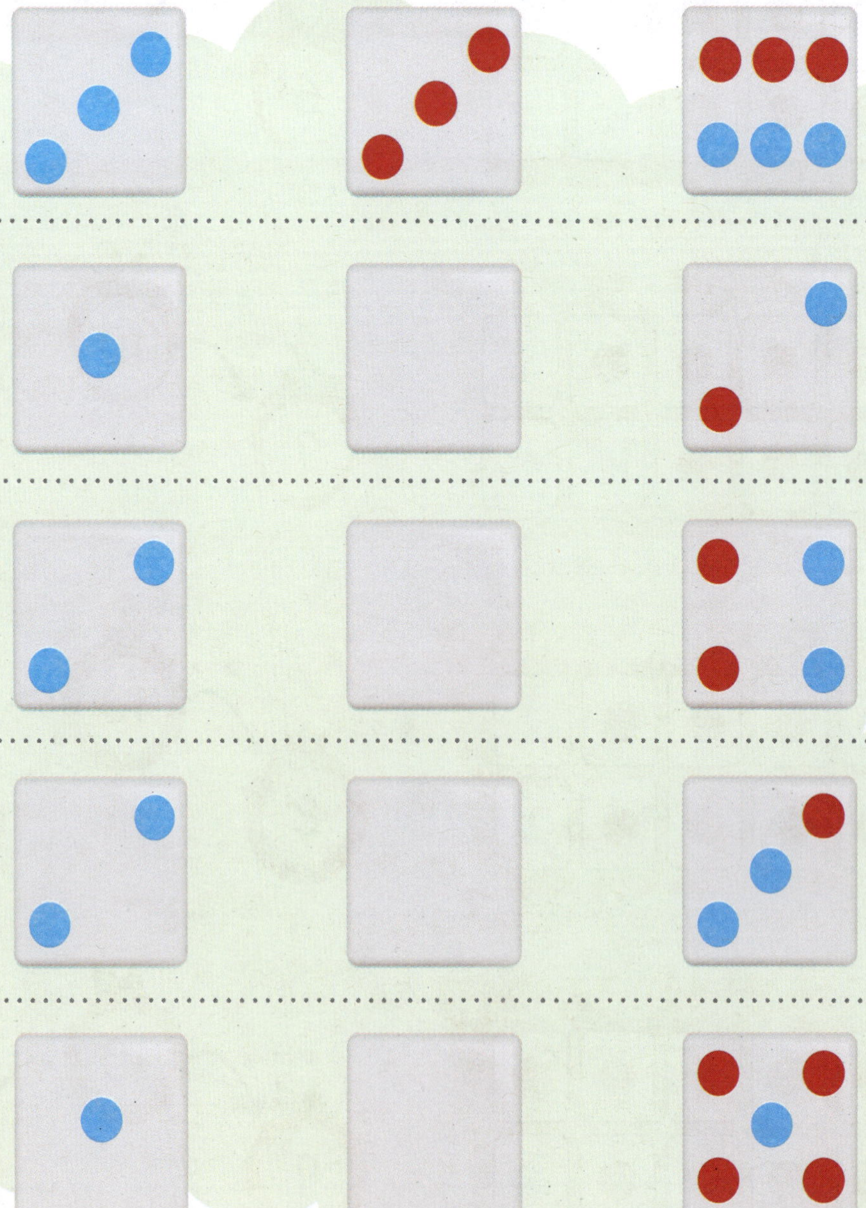

Fill in the missing numbers. One has been done for you.

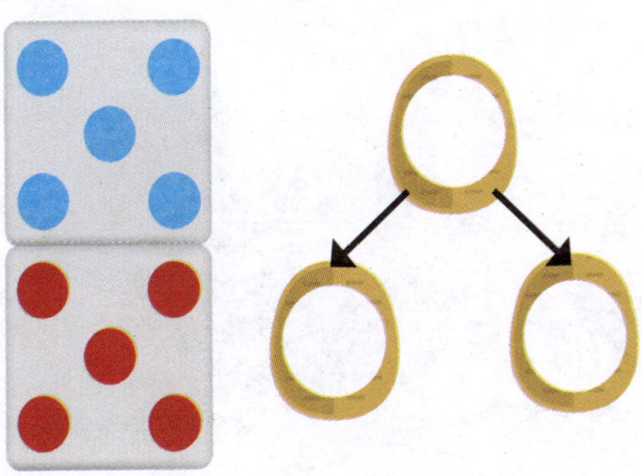

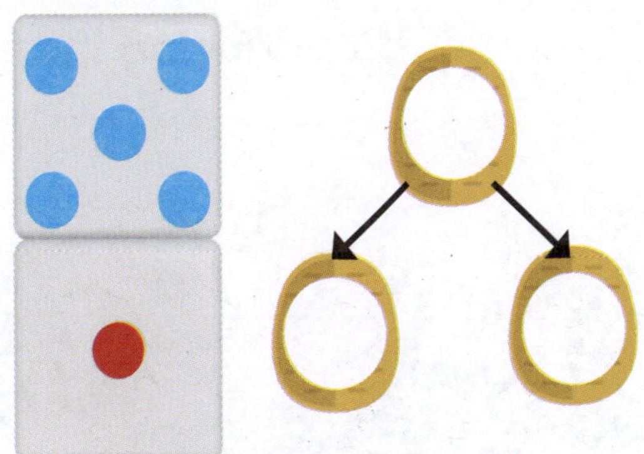

Dancing Butterflies

Sing.

One butterfly with nothing to do
Along came 1 more and now there are two.

Two little butterflies flying free.
Along came 1 more and now there are three.

Three little butterflies flying high and low
Along came 1 more and now there are four.

Four little butterflies touching the sky
Along came 1 more and now there are five.

Five little butterflies play in the garden,
Singing in the rain, dancing in the sun.

Draw 1 more object. Count and write the total number.

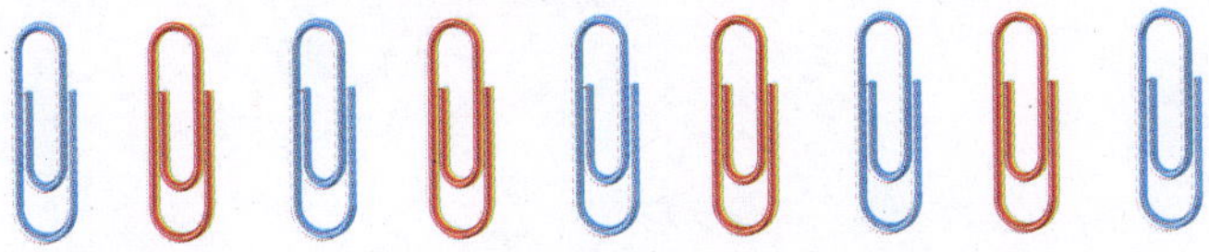

Addition

Count and draw. One has been done for you.

 and  make 

 and  make

 and  make

 and  make

 and make

 and  make

Count and add. One has been done for you.

$$\begin{array}{r} 2 \\ +\ 2 \\ \hline 4 \end{array}$$

$$\begin{array}{r} 1 \\ +\ 3 \\ \hline \end{array}$$

$$\begin{array}{r} 2 \\ +\ 4 \\ \hline \end{array}$$

$$\begin{array}{r} 3 \\ +\ 3 \\ \hline \end{array}$$

$$\begin{array}{r} 5 \\ +\ 4 \\ \hline \end{array}$$

$$\begin{array}{r} 4 \\ +\ 3 \\ \hline \end{array}$$

Five Little Chicks

Sing.

5 little chicks inside 5 eggs,
All of them peck away at the shells.
Out comes 1 and flies away,
4 little chicks still peck away.

4 little chicks inside 4 eggs,
All of them peck away at the shells.
Out comes 1 and flies away,
3 little chicks still peck away.

Continue the rhyme with 3 chicks and 2 chicks, till 1 chick is left.

1 little chick inside 1 egg,
All it does is peck away at the shells.
Out it comes and flies away,
All the chicks fly through the day.

Cross out (✗) 1 object. Count and write the total number.

Subtraction

Cross out (✗). Count and draw. One has been done for you.

10 - 5 =  5

9 - 6 = ☐

6 - 4 = ☐

8 - 7 = ☐

5 - 2 = ☐

Cross out (✗), count and write. One has been done for you.

6
- 2
4

5
- 3

4
- 1

7
- 4

9
- 5

7
- 6

I Know Addition and Subtraction

Count, draw and write.

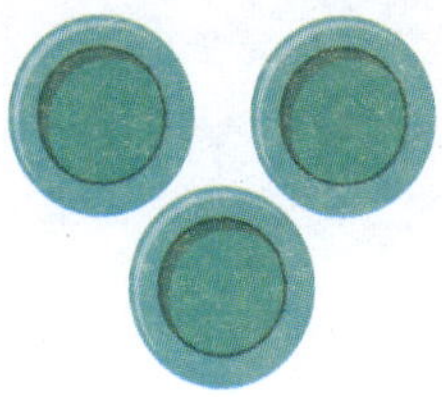 and 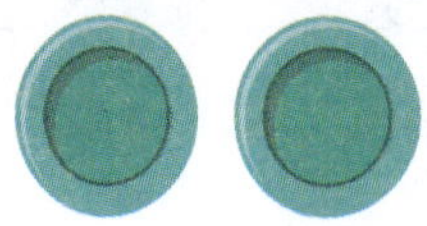 = ☐

 and  = ☐

 and  = ☐

 and  = ☐

 and  = ☐

Cross out (✗), Count and write. One has been done for you.

$$\begin{array}{r} 8 \\ -\ 2 \\ \hline 6 \end{array}$$

$$\begin{array}{r} 4 \\ -\ 3 \\ \hline \end{array}$$

$$\begin{array}{r} 7 \\ -\ 1 \\ \hline \end{array}$$

$$\begin{array}{r} 8 \\ -\ 5 \\ \hline \end{array}$$

$$\begin{array}{r} 6 \\ -\ 5 \\ \hline \end{array}$$

$$\begin{array}{r} 9 \\ -\ 6 \\ \hline \end{array}$$

I Can Count Till 100

Play this game with 3 friends. Each of you select a colour. Roll the dice. Colour every number you land on. The person to reach 100 first is the winner. Have fun!

RED

BLUE

GREEN

START

1	2	3	4	5	6	7	8	9	10
11	12	13	14	15	16	17	18	19	20
21	22	23	24	25	26	27	28	29	30
31	32	33	34	35	36	37	38	39	40
41	42	43	44	45	46	47	48	49	50
51	52	53	54	55	56	57	58	59	60
61	62	63	64	65	66	67	68	69	70
71	72	73	74	75	76	77	78	79	80
81	82	83	84	85	86	87	88	89	90
91	92	93	94	95	96	97	98	99	100

FINISH